FENCES

for
Pasture & Garden

BY GAIL DAMEROW

STOREY
BOOKS

In memory of Carolyn, who started me thinking seriously about fences because she wouldn't stay put.

Cover design by Carol Jessop
Cover photographs by Craig Blouin (f/STOP PICTURES) *upper left;* Paul O. Boisvert *upper right;* David M. Stone (PHOTO/NATS) *bottom*
Text design and production by Carol Jessop
Edited by Kimberley Foster
Line drawings by Carl Kirkpatrick, except those by Allan Damerow on pages 6 and 18 (T-post clip)
Indexed by Gail Damerow

Printed in the United States by Vicks Lithograph & Printing Corporation
20 19 18 17

Library of Congress Cataloging-in-Publication Data

Damerow, Gail.
    Fences for pasture and garden / by Gail Damerow.
        p.   cm.

    Includes bibliographical references and index.
    ISBN 0-88266-754-8 (hc) — ISBN 0-88266-753-X (pb)
        1. Fences—Design and construction.   I. Title.
TH4965.D35   1992
    631.2'7—dc20   91-55486
                                                                            CIP

CONTENTS

INTRODUCTION

▼▲▼▲▼▲▼▲▼▲▼▲▼▲▼▲▼▲▼▲▼▲▼▲▼▲▼▲▼▲▼▲▼

No matter what project my husband and I decide to tackle, it always seems to come down to the same thing: fences. Fences to keep groundhogs and cottontails out of our garden. Fences to keep our dairy goats away from our fruit trees. Fences to keep our chickens in and the neighbor's dogs out. There was even a time when we had to fence in the house to keep our "watch" geese from taking over the basement.

As we quickly learned, installing a fence is a major investment in time and money. Buying materials and putting up a fence, though, is only half of what's involved — the other half is making sure you're building the right fence for the job.

The more we researched the subject, the more discouraged we became about selecting the right fence for our purposes, our terrain, our climate, and our budget. Specific information is hard to come by. What little is available is confusing and fragmented. Even wordy mail-order catalogs, with their inadequate product descriptions and fantastic claims of superiority, make little sense.

One thing has become clear to us — fence technology has changed enormously over the past two decades. However, the information has been slow to spread beyond the bounds of select experts, despite the highly touted information age into which we've supposedly entered.

As our research progressed and the full range of possibilities began to unfold, it occurred to me that the data we were amassing would be as valuable to others as it has been to us. So I've put it all together in this book, including not only information on how to select the right fence for the job but also on how to install it properly so it does the job right.

Selecting the best fence isn't always easy, since each system has its own inherent strengths and weaknesses. Maybe you'll find, as we did, that a combination works best for you. At our place, for example, we erected a post and plank fence at our driveway entry because it's impressive and picturesque, especially with purple irises surrounding the posts and shocking-pink roses climbing the rails. We use electric fences along the property line as a sure but inexpensive way to keep our animals in and protect them from marauding dogs and coyotes. We use stock panels to control motivated livestock during breeding season, woven wire to confine our chickens, and portable electroplastic netting to pasture young animals in spring. We've learned, you see, that there is no one "right" fence — only the right fence for the job at hand.

One of our current jobs is trellising grape and berry vines. When we discovered high-tensile wire, we thought it would be perfect for solving our sagging trellis wire problem. It was. But we didn't really discover anything new; high-tensile wire has long been used for trellises in Australia and New Zealand, and was introduced to commercial vineyards here ten years before we "discovered" it.

This book, then, describes not only barriers that deter animals from getting in or getting out, but also trellises that confine permanent plantings such as grapes, brambles, and espaliered fruit trees. It also covers fences designed to control blowing sand or snow. At the back of the book is a list of sources for fencing materials and a glossary explaining the specialized language used by contractors and suppliers.

I would like to thank my husband, Allan, for

helping with the charts and illustrations and, throughout my research and writing, for seeing that I never ran out of my primary source of fuel, hot tea.

Also a great big "thank you" goes to Bill McCamman of New Zealand Fence Systems and John McCall of Lifetime Fence for reviewing the entire manuscript and keeping me on the straight and narrow;

to the "snowman," Bob Jairell of The Rocky Mountain Forest and Range Experiment Station in Laramie, Wyoming, for bringing me up to date on snow fences;

to Bill Sites of the U.S. Forest Service in Asheville, North Carolina, for filling me in on the latest in wood and wood preservatives;

to Randy Hussung of Seymour Manufacturing for enlightening me on post hole diggers;

to George and Jennie Ivey, novice fence builders, for ferreting out potentially confusing passages so I could fix them;

and finally, to Stan Potratz of Premier Fence, for inspiring me to write this book in the first place.

CHAPTER 1

PLANNING YOUR FENCE

▽▲▽▲▽▲▽▲▽▲▽▲▽▲▽▲▽▲▽▲▽▲▽▲

"**B**uild all your fences horse high, pig tight, and bull strong," the old saying goes. But, with all the different options, how do you decide which kind of fence to build in the first place? The possibilities are nearly limitless, and no one else's fencing needs are exactly like yours.

Your purpose in putting the fence up will, of course, automatically narrow your choices. Other factors include the lay of the land, your budget, the preferences and prejudices that dictate your sense of aesthetics, zoning restrictions, and whether your fence will be permanent or movable.

SELECTION

Exactly which fencing system will do the job depends on what kind of animals are involved, how large they are, how crowded they are, and how determined they are to get in or out. Among livestock, dairy cows and beef cattle are the easiest to contain. Horses are only a little less so. Next come pigs, sheep, goats, and game animals, in that order. Fowl aren't hard to confine, provided they can't slip through the fence or fly over it.

Knowing the habits of your animals will help you select the right fence to keep them in — are they climbers, crawlers, diggers, chewers, back-rubbers, or head butters? Don't overlook seasonal characteristics such as the ability of baby animals to slip through an otherwise sturdy fence or the propensity of breeding-age stock to go nuts in season.

More difficult than keeping stock in can be keeping predators out, a task that includes both preventing hungry coyotes from getting at the sheep and deterring groundhogs from the pea patch. Here again, knowing the habits of animals helps enormously. Coyotes, for example, tend to be less adventuresome than dogs about getting through a fence, but once they have a taste of what's on the other side you will have a devil of time keeping them from coming back.

Just how persistent predators are likely to be depends on how hard they have to compete for food and water outside the fenced area. Foxes will be more attracted to the henhouse after they have decimated the local rodent population. In droughty areas, wildlife may break in simply to get a drink. A fence crossing a migration trail or separating wildlife from their traditional breeding grounds offers a powerful incentive for breaking through. Even if the animal's original goal wasn't to get at your livestock, your fence has been breached with potentially disastrous results.

Options

The dizzying array of permanent fencing systems suitable for pasture or garden can be broken down into four basic styles: nonelectrified parallel strands of wire, electrified parallel strands, woven, and rail. In evaluating these systems, the main features to consider are function, appearance, ease of installation, sturdiness, safety, life expectancy, required maintenance, and cost.

At one time, barbed wire made the cheapest kind of fence and a lot of people still think it's the only affordable game in town. But it is only suitable for confining cattle on open range. Unlike more curious or spunky forms of livestock, bovines won't go bonkers if they get tangled in barbed wire. But the barbs will damage horses, snag the wool of sheep, tear up the udders and ears of goats, and rip your arm open if you're not careful. In some places, using it is illegal.

High-tensile smooth wire is a safer, more attractive alternative. Strand for strand, it costs about the same as barbed wire but is less costly than woven wire. Built right, a high-tension fence will keep any large animal in or out. Used as trellising, it is strong enough to support the heaviest fruit-laden vine.

Electrify high-tensile wire and you can both reduce the cost of your fence and increase the range of animals it will deter. Not quite as sturdy, but somewhat easier to build, is an electrified low-tension fence. Electrifiable wire wrapped on reels can be used for temporary perimeter fencing or to cross-fence large fields.

A well-built woven fence, though not cheap, will keep nearly anything in or out. No other kind of fence beats it as a physical barrier. It's the traditional choice for sheep, goats, poultry, and gardens. The most expensive but sturdiest options are made of galvanized or plasticized metal, but for some purposes relatively inexpensive plastic mesh will do just fine.

Rail, the most expensive and the most elegant of all fences, is popular for horses, especially young, frisky ones that easily get tangled in woven or barbed wire. A good rail fence can withstand the kicking, charging, and chewing that horses inflict on fences, and in return is less likely to injure the horse. Once constructed primarily of wood, rail fences are now also made of rot-resistant vinyl, steel, and even concrete. An electrified version can be used where horses are pastured with other stock or where predator control is necessary.

Which of these systems will work best for you depends, in part, on your terrain. Rigid options like rail and woven are more suitable for level or gently rolling land, while wire strand fences adapt more easily to rough or hilly country.

Another consideration is how much pressure the fence is likely to experience from animals and machinery. Livestock crowded into a barnyard lot or corral for feeding or loading exert more pressure on a fence than stock in an open field. Tractors and trucks used to harvest crops exert pressure if they rub or bump against the fence. For high-pressure situations, your best choice is a sturdy fence that has a bit of give and easily bounces back. Cable and high-tension fences make two good possibilities.

Local zoning laws may further restrict your choice. Before making any decisions, check with your city or county planning office to find out if there are any height restrictions or other regulations pertaining to fence design or construction. If you need a permit to build a fence, your design will likely be reviewed and approved or modified before the permit is issued.

If you want to confine a variety of animals in one area, if the purpose of your fence is subject to change, or if you need to strengthen an existing fence, a combination fence may be your best option. Popular combinations include plastic mesh fastened to rails, electrified wire augmenting rails or woven wire, and barbed wire strung beneath or above a woven fence.

Whatever system you choose, base your decision on the worst possible situation. Coyotes in the east and west, wolves in the north, and dogs everywhere create an ever-increasing threat to livestock. Canines are especially a problem in spring, when they run in packs and when most livestock have young. In some areas, increasing populations of deer, elk, or moose are beginning to pose a serious challenge for gardeners.

Just because you haven't had a problem in the past doesn't mean you can count on good luck in the future. Building a secure fence today will cost you less than reinforcing an inadequate fence tomorrow, especially if you include the cost and aggravation of losing valuable animals or plantings.

Cost

In the real world, cost is usually the deciding factor in selecting a fence. Unfortunately, there's no such thing as a "cheap" fence, in the sense of low in cost. A truly cheap fence is one that offers good insurance, not one that's initially cheap to put up. Selecting materials of lesser quality will only shorten the useful life of your fence and increase both maintenance and the cost of repairs. Some of the most expensive fencing systems happen also to be the longest lasting, which makes them the best buys in terms of cost per year of use. Like the man says, "You can pay me now, or you can pay me later."

When you find out how much it takes to finance a good fence these days, you may fall into a dead faint. Before you decide to take out a second mortgage, do some price comparing between various sources. Because prices change so quickly and vary so much from place to place, it doesn't make much sense to list specific costs here. To find out how much your fence will cost, study mail-order sources (some are listed in the appendix) and visit local merchants.

Beforehand, prepare a map of the area you want to fence, with all the appropriate measurements. If you're fencing a big area, especially one that is far from level, you might start with a topographic map purchased for a few dollars from the department of geology in your state's capital. For a few more dollars, a good print shop will magnify the part showing your property. If you're fencing a smallish area or one that is fairly level, use a sheet of ordinary graph paper to keep the scale accurate.

On your map, mark the locations of corner posts and gate posts. Indicate where any significant rises or dips occur along the fence line. Make a mark for each line post, spaced according to the guidelines for the kind of fence you plan to build.

Your map will help you prepare a shopping list, or bill of material, listing how many and what kind of posts you will need (as described in chapter 2), fasteners to attach fencing material to the posts (also chapter 2), anchor assemblies (chapter 3), gates and gate hardware (chapter 13), and lightning protection (chapters 5 and 8). If your fence will be electrified, you will also need an energizer and related components (chapters 7 and 8).

Add up the total length of the fence line on your map to calculate how much fencing material you will need. For rails or wire strands, multiply the total length of the fence by the number of rails or strands you plan. If, for example, your fence line measures 1,000 feet (305 m) and you want a three-rail fence, you will need 3,000 feet (915 m) of rail. Next, determine how the fencing material is sold, whether in 16-foot (5 m) rails, 165-foot (50 m) rolls, or whatever, so you can calculate the total number of *units* you need, rounded up to the nearest whole number.

FENCE SELECTION GUIDE

	Barbed Wire	High-Tension	Low-Tension Electrified	High-Tension Electrified	Woven Wire	Rail
Purpose	cattle	cattle	cattle	cattle	cattle	cattle
			goats	goats	goats	
		hogs	hogs	hogs	hogs	
		horses	horses	horses	horses	horses
		mules	mules	mules	mules	mules
			sheep	sheep	sheep	sheep
			poultry	poultry	poultry	
		corral			corral	corral
		game		game	game	
			garden	garden	garden	
		trellis				
Appearance	poor	good	fair	good	good	excellent
Function	perimeter	perimeter	perimeter	perimeter	perimeter	perimeter
	cross fence		cross fence			
				security	security	
Life (years)	30 to 60	40 to 60	15 to 25	45 to 65	25 to 55	15 to 30
Sturdiness	low	high	low	high	medium	medium
Maintenance	high	low	medium	low	medium	high
Safety	poor	good	good	good	good	good
Predator control	low	medium	medium	high	high	low
Limitations	illegal in some areas	few	illegal in some areas; post signs	illegal in some areas; post signs	few	frequent painting
See chapter	6	6	9	9	11	12

If erecting a fence sounds complicated, you may consider hiring a contractor. Still do the preliminary layout to be sure the contractor knows what you want and to make certain you get what you pay for. If you decide to go ahead on your own, you will find it isn't all that difficult. And you will have the satisfaction of knowing that you have saved as much as 50 percent — the usual fee for contract labor tacked on to the cost of materials.

Some of the suppliers listed in the appendix will happily take time to help you plan your fence. Some even have computer programs that will list every item you need, right down to the last loop cap or twitch stick. Your county extension agent also may be able to supply helpful information. Other sources include landscape designers, fencing contractors (called "fencers" in the trade), and fencing materials suppliers listed in the yellow pages. If you consult someone from whom you don't plan to buy products or installation services, ask in advance how much you will be charged for consultation.

Weight

Fencing materials can be quite bulky and heavy. If you're thinking of doing the whole job yourself, reconsider and try rounding up a partner or two. Weight also can be a factor in figuring the cost of your fence. If you're planning to have materials shipped, make room in your budget for the cost of freight.

If you plan to pick up your materials from a local supplier, your vehicle may not handle all the weight in one trip. Consider splitting up your shopping list. In one trip you might, for example, bring home all your posts and a post driver or earth auger, and go back for the remaining materials after your posts are in. Or you might bring home all the materials you need to put in, say, one-third of your fence at a time.

If you don't mind the extra cost and you can find a willing supplier, you could have everything delivered to your door. On the other hand, if you opt for a system calling for fiberglass posts and lightweight wire, the whole thing just might fit into the trunk of the family sedan.

TOOLS AND SUPPLIES

While you're gathering materials for your fence, include a few simple tools and supplies. You will need the items listed here no matter what kind of fence you install. Other, more specialized tools are discussed in the chapters pertaining to their use.

Before taking any hand tool into the field, paint it a bright color. Green blends in too well with grass and weeds, yellow or international orange disappears in autumn leaves. You will save lots of time otherwise spent hunting for tools on the ground, as well as money to replace those you can't find, if you paint all your tool handles pearly white or electric blue.

A tool caddy comes in handy for organizing hand tools, supplies, and other paraphernalia. For small jobs, you can use one of those colorful plastic tool caddies sold in the household section of any department store. A leather carpenter's apron frees up your hands if you're carrying such things as staples and fence clips. For large jobs, you might use a set of well-labeled coffee cans held in a sturdy carrier equipped with hand grips, of the sort used by folks who deliver milk.

String

String is indispensable for getting posts to line up in a row. Cotton mason's line is available at lumberyards. Nylon cord sold at hardware stores stretches too much and unravels too easily unless you remember to touch the cut ends with a match.

Stakes

You will need lots of stakes for your preliminary layout. On fairly level land or for small runs of fence, 2-foot (60 cm) wooden pegs are fine. On hilly country or for long distances, taller stakes are easier to spot. Fiberglass rods are still easier to see than wood stakes, since from a distance the latter tend to blend into the surrounding countryside unless you paint them white or orange.

Surveyor's Ribbon

Brightly colored plastic surveyor's ribbon has all sorts of fencing uses. You can tie it to the tops of layout stakes to make them easier to see, wrap it around trees to remind you which ones are in the way and have to come out, or use it to warn man and beast of a newly strung electric fence.

Tape Measure

You will need a measuring tape to ensure that posts are evenly spaced and to attach fencing material to the posts at the right heights. A long tape, such as a 100-foot (30 m) reel tape, works best for measuring the distance between posts. To establish the correct height for wire or rails, a 25-foot locking carpenter's tape comes in handy; handier yet for this purpose is a template of the sort described in chapter 5.

Levels

To keep your fence level and plumb, you will need two different kinds of level. One is a long metal carpenter's level used to get your posts vertical (or plumb). Check each post by holding the level against two adjoining faces. When setting end or curve posts off level, tape a wood block of appropriate size to one end of the level and orient the block toward the top when you check the post. If, for example, you want posts to lean 2 inches (5 cm) out of plumb, the block should be 2 inches (5 cm) thick. The other kind of level is a line level you hook onto a string stretched between posts; this level helps you get your fencing material fully horizontal or parallel to grade, as the case may be.

Safety Clothing

You will need a few items for personal protection. One is a pair of safety glasses or an eye shield in case wire under tension gets away from you, or splinters fly when you drive in a wood or metal post. Another is a pair of heavy leather gloves, especially if you will be working with barbed wire, cable, or wood, but helpful for any fencing project.

If you will be stringing a lot of barbed wire, you might want a pair of chaps to protect your jeans and legs. You should definitely wear boots and tough, close-fitting clothing, particularly if you will be using power equipment to drill holes or drive posts.

LAYOUT

To map your fence, sooner or later you will have to go out in the field and measure. Begin by driving stakes into the ground where your corners and gates will go. This preliminary step serves as an aid to measuring; it also helps you decide if this is *really* where you want the fence to be. It may show you that you will be putting a gate in the wrong place or that you need additional gates. And it establishes exactly where you will need to install each post.

From the standpoint of strength and ease of installation, the ideal fence consists of straight runs joined at right-angle corners. In terms of grazing efficiency, a square paddock is more evenly grazed than a long, narrow one, which tends to get overgrazed at the front and undergrazed at the back.

From a maintenance standpoint, sweeping curves are easier to mow than corners. Curves, however, may require extra bracing, which will add to your labor and cost. Whether or not to curve the corners of your fence depends somewhat on the size and shape of the area you are fencing — curved corners will reduce the total inside area.

If you are enclosing animals that get their drinking water from a stream, leave only a narrow corridor to the water. Fence out as much highly erodible stream bank as possible so vegetation won't be eaten or trampled. If you practice controlled grazing and you need a lane to get animals back and forth from water or to drive them from one pasture to the next, minimize erosion by arranging the path along a natural ridge or contour line.

When building a boundary fence, know where your property line is — this may entail hiring a surveyor. Local setback restrictions may dictate how close to your property line you can place your fence. You will also want to check with your highway commission and check your deed to ensure you won't be putting the fence inside a right-of-way or across an easement. If you are running the fence along a road or right of way, always measure from the *center* of the roadway.

If you plan to put a fence along your property line, your neighbor may be willing to share its cost and maintenance. Get an agreement in writing, detailing all the specifics. Where long-term maintenance is involved, record any agreement that allows you to enter adjoining property to repair your fence. You and your neighbor may be best friends now, but tomorrow some old grouch may move in next door.

In the event you can't get a written agreement, build your fence sufficiently inside your property

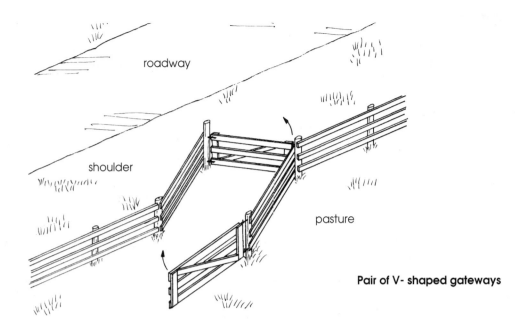

roadway

shoulder

pasture

Pair of V- shaped gateways

line so you can mow and otherwise maintain both sides. At the very least, allow enough setback so concrete footers and other protruding parts won't encroach on the neighbor's land. Some future challenge to the placement of your fence could end up in a costly court battle and the subsequent moving of your fence.

Before building any kind of metal fence beneath power lines, seek safety advice from the power company. Make sure, too, that your proposed fence won't interfere with any underground structures such as a septic tank and its leach lines, and check with local utility companies about underground cables and water or gas pipes that may require the rerouting of your fence. A one-call service to help you contact all the appropriate utilities is listed in the appendix.

Gate Placement

Discovering that your gates are all in the wrong places can be annoying at best and downright inconvenient at worst. Moving gate posts is time-consuming and expensive, so give serious thought beforehand to where you will put your gates. Place them in well-drained areas to avoid muddy conditions. Keep them out of the path of erosion, since through traffic will only make matters worse.

Take into consideration your normal patterns of movement and put gates where they will be the most convenient. If you are fencing a pasture or large garden, a gate near the corner will encourage vehicle or foot traffic to move along the fence instead of cutting down the middle. If you are confining livestock, a corner gate lets you drive animals along the fence and out.

A gate that opens onto a roadway should be set far enough back so you can pull your vehicle off the road while you get out to open the gate. A generous setback is especially important on a narrow road with little or no shoulder.

As an alternative, to avoid obstructing the right-of-way, construct a V-shaped gateway, with the gate opening in the direction you normally travel. If you keep the sort of animals that hang around a gate and try to scoot through every chance they get, add a second V-shaped gateway that works something like an airlock. You go through the first gate and close it, then open the second gate and go out.

If you have fields on both sides of a roadway, you may find it convenient to place gates opposite each other. On a private road or driveway, pairs of opposite gates can be arranged so that when both are opened, livestock can be channeled directly from one field into the next with no opportunity to stray down the road.

For a full discussion of gate sizes, styles, and construction, see chapter 13, "Gates."

CLEARING

Before starting your layout, you may have to clear along the fence line. Clearing may entail mowing weeds or it may involve cutting down trees and smoothing the grade. Grading out dips and bumps will simplify fence installation and may even let you reduce the number of posts you need. A cleared fence line makes it easier to maintain the fence once it's up. A wide clearing renders the fence more visible to wildlife on the outside, to livestock on the inside, and to any two-legged critters who may wander by.

A well-cleared fence line also increases the longevity of a fence. Encroaching weeds and brush hold moisture that encourages a wooden fence to rot, a metal one to rust, and an electric wire to short out. If wooded or brushy land adjoins the fenced area, a 10-foot-wide (3 m) clearing provides a good fire barrier, minimizing the chance that a fire will ruin your fence.

If your proposed fence line is highly overgrown or has lots of dips and valleys, it is well worth having a bulldozer come in to clear and level. For a short fence, you can do the same job yourself with a mattock, shovel, and rake, but the work is both time-consuming and back-breaking. For a fence of any substantial length, a dozer can do the same job in hours that may take you weeks by hand.

If you don't have many large trees, boulders, or huge piles of soil to contend with, use a tractor to pull up small trees, to push rocks out of the way, and

to smooth out the land. Clearing seriously overgrown or hilly land with a tractor can be quite dangerous, though, so don't attempt it unless you are highly experienced.

To establish a fence line in an area that is so overgrown you can't see where you are going, set tall poles where you want your corners to be, run a string between the poles, and clear beneath the string. If the distance between the corners is too far to make running a string practical, set a tall pole at the most overgrown corner and place a 5-foot (150 cm) stake at an adjoining corner. Stand behind the stake, facing the pole, and have a partner align a second pole as far into the undergrowth as possible between them. Clear a swath between the stake and the first pole. Replace the pole with the stake and repeat the procedure, continuing until you have cleared all the way to the corner. Where overgrowth is extremely heavy, you may have to clear an approximate fence line, then go back later and straighten it up.

If you are clearing along an old, dilapidated fence, never be tempted to shore up any part of it. The job could end up costing more than putting up a new fence from scratch. You might, however, be able to salvage some of the old posts for reuse. Instructions for pulling up posts without damaging them appear in chapter 4.

Don't be tempted, either, to leave in the old fence and run your new one parallel to it. The derelict fence will look unsightly next to your fine, new

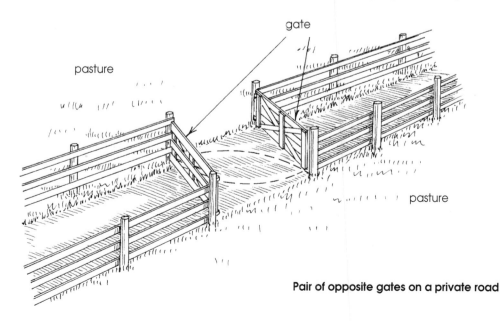

Pair of opposite gates on a private road

fence. Besides, weeds growing around the old fence will make it a hazard to anyone who later tries to clear, or to animals that might get tangled in it.

When taking down an old fence, begin by removing the old wire and fasteners from the posts. Make sure you don't drop any staples. Lost staples invariably end up embedded in a truck or tractor tire when you least have time to fix a flat. Besides, some animals have the habit of eating dropped staples, thereby contracting an often fatal malady called "hardware disease."

Take care, too, to pick up all broken-off bits of wire. They can wrap themselves around a mower blade or, worse, around livestock. One fellow learned this the hard way after carelessly discarding a short length of wire in a field while repairing his fence. He came back later to find the wire had strangled a newborn lamb.

Maintenance

Maintaining a fence is largely a matter of keeping the fence line cleared so you can easily walk or ride along it to inspect and make repairs. How often you need to check your fence depends on what kind of fence you have. On open land, you might inspect a sturdy rail or woven wire fence only as often as you mow. An electric fence in wooded country, on the other hand, should be inspected daily — which takes mere seconds using the handy little fence-tester described in chapter 7.

If you take time to cut down weeds in the spring when vegetative growth is highest, they will be much easier to keep back throughout the rest of the year. You could, if you wish, spray an herbicide along the fence, but select one that is not toxic to livestock and avoid corrosive glyphospate around fences made of galvanized wire. If money is no object, you might use some permanent means of weed control, like pouring a concrete footer or running vinyl landscaping strips along the fence line.

Even if you don't use concrete or vinyl for the whole fence, you might wish to use one or the other around gate posts to keep them looking neat, and in odd-angle corners that will otherwise be hard to keep clear of weeds. Whatever method you choose for controlling weed growth, don't ever consider fire — you will weaken your posts and melt the protective coating off the fencing material.

You may be tempted to let brush grow up along your fence as a wildlife habitat, but if you do, you will not only shorten the life of your fence but will need a dozer to clear out the mess when time comes to build a replacement. If you simply *must* maintain a hedgerow, build your fence 10 feet (3 m) away and parallel to it. Your field will be smaller and any crops you grow will be inundated with wind-blown weed seeds, but you can rest assured that you have done your part for the birds and bunnies, and your fence will be free to live out its full, natural life.

Clearing a fence line

CHAPTER 2 ▼▲▼▲▼▲▼▲▼▲▼▲

ALL ABOUT POSTS

▼▲▼▲▼▲▼▲▼▲▼▲▼▲▼▲▼▲▼▲▼▲

A fence is only as strong as its posts, making posts the most important part of any fence. They're also the most expensive part. It makes good sense, therefore, to take time and care in selecting them. The kind of posts you choose will depend, in large part, on what kind of fence you build. Most fences require at least two different kinds of post, according to their role within the fence.

ANCHOR AND LINE POSTS

You will need stout posts at key spots such as corners, curves, dips, rises, and gates. Posts in these positions are called "anchor" posts since they anchor the fence down, giving it strength and stability. Anchor posts are generally larger in diameter than line posts and they are longer so they can be set deeper into the ground.

The evenly spaced posts between the anchor posts are called "line" posts. They need not be as strong as anchor posts since they incur much less stress. Their primary purpose is to secure the fencing material. The taller your fence, the longer your line posts should be, not just to accommodate the higher fence but also so that you can set them deeper into the ground to support the weight of the fence material.

If your soil is sandy or moist, if you keep animals in close confinement, or if you space your line posts farther apart than usual, you will need stouter line posts than otherwise. If you are stringing a wire strand fence, you can get by with fewer line posts by using lightweight spacers between them.

Both anchor posts and line posts should be as straight as possible. Besides looking bad, crooked posts place extra strain on fencing material.

Anchor posts, line posts, and spacers may be purchased at farm stores, building supply outlets, and local or mail-order fence suppliers.

Not incidentally, using trees as fence posts is a bad idea for several reasons. First, trees attract lightning, which can seriously damage your fence. Second, as the tree grows, the fencing material will grow into the trunk, damaging both the fence and the tree. Finally, some future woodsman may not know the tree has been imbedded with wire, staples, or nails, and dire consequences to life or limb may result when the saw hits metal. If you insist on attaching fencing material to a tree, protect the tree with a 2x4 (5x10 cm) and attach the fencing material to that.

TYPES OF POST

Fence posts are made of wood, steel, concrete, stone, plastic, and fiberglass. Each has advantages and disadvantages. Wood makes strong but fairly expensive posts that deteriorate more rapidly than most other kinds. Steel is longer lasting than wood, but not as strong. Concrete posts are durable, but not widely available. Stone posts are as sturdy as concrete but are even more difficult to find, and can be quite expensive. Plastic is inexpensive but doesn't hold up well. Fiberglass is strong, resilient, and relatively inexpensive.

Wood Posts

Originally all fence posts were made of wood. Wood posts remain popular because they come in a wide range of diameters, heights, and such inherent characteristics as hardness, durability, and appearance. Wood is also versatile, and can be used in acid soils where steel posts would corrode. Wood posts have a natural yet sturdy look, and that sturdiness isn't all show.

Stout wood posts make good anchors for highly tensioned fences and those designed to withstand constant animal pressure. Unless you cut your own, though, they can be expensive, especially when you consider how quickly some decay.

When it comes to wood rot, looks can be deceiving. Don't toss out old posts with rotting bottoms too quickly. Stick a knife into the worst-looking spots. If you hit a solid core, it is likely that enough sound heartwood remains to make the post usable for a while longer.

If you have sufficient land and you plan to stay put, you may want to grow your own replacement posts. Since the most durable posts take the longest time to grow, consider this project to be a long-term investment.

When the time comes to harvest, select trees that are as close as possible to the diameter you need. Seek out those with their lowest branches higher than you want your posts to be, and you will get straight, strong posts with few knots and little taper.

The ideal time to cut trees for posts is spring, just as new growth begins. The bark loosens then, making a post easier to peel. At other times of year, stripping bark is hard work and you may not get it all off. Bark remaining on a post holds moisture and provides a hiding place for insects, thereby decreasing the post's longevity.

To peel bark, first anchor the post so it won't roll. You might, for example, cradle it between stout nails hammered securely into the tops of two sawhorses. Loosen the bark by pounding it with a sledge hammer. Then pull an ax or drawknife down the length of the log in long, smooth strokes, peeling off a strip at a time. When you finish one side, turn the log and peel the other side. With some species, after peeling the first strip you can pry away the bark and remove it in chunks.

Whether your grow your own posts or buy them, the accompanying chart offers some guidelines regarding the approximate longevity of various kinds of wood. As you might surmise from this chart, it is impossible to tell exactly how long a wood post will last. Since wood is a natural product, its durability varies considerably and is influenced by local temperatures, rainfall, and humidity.

Within any species, the posts that last longest are those with their bark removed, that were dried before being set into the ground, and that consist mainly of heartwood. Heartwood, the darker wood at the center, resists decay. Sapwood, the lighter wood between the bark and the heartwood, deteriorates rapidly.

Shapes and Sizes

Wood posts may be left round like the logs they came from, split into half-rounds (cut or split in half lengthwise), faced (having a slice taken off one side to remove sapwood and create a hard, flat nailing surface), or squared (milled to create four flat sides). Round posts are stronger than milled posts and are used mainly for field fences. Half-round posts, faced posts, and square posts are used primarily for board fences.

The softer the wood, the larger in diameter the post must be for equivalent strength. Round posts range upwards in diameter from 2½ inches (65 mm), measured at the smaller end. Posts that small may be used for straight runs of field fence on level terrain, but posts of at least 3½-inch (90 mm) diameter make a stronger fence. For corrals and other places of close confinement, or in wet or sandy soil, line posts should be at least 4 inches (100 mm) in diameter.

Even if your line posts are of some other material, wooden corner and gate posts will give your fence extra strength. Anchor posts should have a diameter of at least 5 inches (130 mm). A gate post supporting substantial weight should be at least 8 inches (200 mm) in diameter. Wood anchor posts should be as straight as possible, free of splits, and either round durable hardwood or pressure-treated milled lumber.

The size of a milled post is indicated by the same series of three numbers used for any lumber, for example 4x4x8. The last number specifies the post's height, in this case 8 feet (240 cm). The other two numbers give the dimensions in inches of the cross-section (here, 4 inches or 100 mm square). The first two numbers actually indicate the "nominal" size of the rough-sawn post before it is dried and planed smooth. The actual size of a 4x4 after milling is about 3½ inches by 3½ inches (90 by 90 mm). (Since the spacing between posts is measured from the center of one post to the center of the next, you don't really need to worry about true dimensions.) Milled line posts may be 4x4s, anchor posts at least 6x6s.

Wood posts generally come in lengths from 5½ to 8 feet (165 to 240 cm), although you can get them as long as 12 feet (360 cm) for fencing game animals. If you can't get the length you want, get the

LONGEVITY OF FENCE POSTS

(round, 5" to 6" diameter)

Wood Species	Notes	Life Expectancy in Years	
		Untreated	Treated
Ash		3–7	10–15
Ash, black		2–10	
Aspen		2–7	15–20
Australian ironbark	marketed as "Insultimber"; chemical treatment unnecessary	40+	
Baldcypress, old gr.	naturally termite resistant	7–15	20–25
Baldcypress, young gr.	naturally termite resistant	2–7	
Basswood		2–7	15–20
Beech		2–7	15
Birch		2–7	10–20
Boxelder		2–7	16–20
Butternut		2–10	15–20
Catalpa		7–20	15–25
Cedar, red	inferior nail-holding capacity	7–20	21–30
Cedar, white	inferior nail-holding capacity	10–20	21–30
Cherry, black		5–10	
Chestnut	inferior nail-holding capacity		
Cottonwood		2–7	10–15
Cypress		5–15	
Elm		2–7	15
Elm, slippery		5–15	
Fir, balsam		2–7	10–15
Fir, douglas		2–7	15–18
Hackberry		2–7	10–17
Hemlock, eastern		2–10	10–25
Hickory		2–7	15–20
Larch		2–7	10–20
Locust, black	chemical treatment unnecessary	15–40	
Locust, honey		2–7	
Maple		2–7	15–20
Mulberry, red		7–25	15–30
Oak, black		5	
Oak, red		2–7	15–20
Oak, white		5–15	15–20
Osage orange	chemical treatment unnecessary	15–40	
Pine	moderate nail-holding capacity	2–7	20–30
Pine, southern yellow		2–7	20–35
Poplar, yellow	inferior nail-holding capacity	2–7	20–26
Redwood	naturally termite resistant	7–15	20–30
Sassafras		7–15	20–25
Spruce	moderate nail-holding capacity	2–7	10–20
Sweetbay		2–7	10–20
Sweetgum		2–7	20–30
Sycamore		2–7	20–26
Tamarack		2–10	15–20
Tupelo, black		2–7	15–20
Willow		2–7	15–20

next longer size and either set the posts deeper or cut off the excess. As a rule of thumb, posts should be half-again as long as the fence is high, since about one-third gets buried in the ground (allow even more depth if your area is subject to frost). Smaller posts may be sharpened with a hatchet into a pencil point at one end so they can be driven in with a maul or power driver. Larger posts may have to be hand set in dug holes, especially if the soil is hard or rocky.

Some fencers have the mistaken idea that if they round off or slope their post tops, the posts will more readily shed rainwater and therefore rot less quickly. The fact is, decay begins at or below ground level. Furthermore, if you bevel your post tops, you will increase the surface area and the posts will absorb more moisture, not less.

Timber Quality

Wood deterioration is caused by fungi and insects such as termites, beetles, and carpenter ants. Fungi and insects need four things for survival: moisture, warmth, oxygen, and food. Their food is the wood itself, which you can eliminate by using posts of steel, concrete, or fiberglass. Assuming you prefer posts of wood, you'd be hard-pressed to control the amount of moisture, warmth, and oxygen in them. The exception, of course, is posts set entirely underwater. Even though they are full of moisture, they won't rot because there's not enough oxygen.

You're left, then, with eliminating the food value of the wood and thus the fungi and insects. Nature does just that by imbuing heartwood with natural toxins, which is why wood posts with a high heartwood content last the longest. The heartwood of two of this country's commercial species — redwood and baldycypress — is highly resistant to decay, termites, and weathering. Cedar also rates high.

Sadly, our forests have been wastefully exploited over the past two centuries, and now we're paying the price. Much of the lumber you find on the market today is cut from soft, new growth that lacks a substantial amount of heartwood. You can readily identify this less dense wood because its growth rings are relatively far apart. Posts made from it rot rapidly, and fence rails warp easily.

Any post cut from softwood or sapwood has a short lifespan, usually three years or less. That's not very long, considering all the money, time, and effort that goes into building a fence.

Pressure Treatment

To compensate for the scarcity of woods with a high heartwood content, a process has been developed to increase the longevity of softwoods (as well as their price). Chemical preservatives are forced under pressure into wood cells, where they become permanently locked in and help the wood resist attack by wood-destroying organisms.

The useful life of a post thus treated can be extended to as much as fifty years. Pressure treatment does require the use of noxious chemicals but, in one of those ecological trade-offs we are all learning to live with, it reduces the number of trees that have to be cut down to replace rotted wood.

Three major kinds of wood preservative are currently used in pressure treatment: creosote, oil-borne, and water-borne. Creosote, the oldest wood preservative, is favored by prairie farmers because treated posts shed water and are resistant to grass fires. After charring at the surface, the posts simply self-extinguish. Creosote is an organic product derived from coal tar. Eventually it will biodegrade, but in the meantime it gives off harmful vapors, particularly after it has been freshly applied. It can kill nearby plants and can be harmful to horses and other animals that crib (chew) or lick wood, a problem you won't have to worry about if you're building an electric fence. Creosote is most often used for posts set in salt water.

Oil-borne preservatives consist of an alphabet soup of chemicals including iodo propynyl butyl carbamate (IPBC), pentachlorophenol (penta), tributyl tinoxide (TBTO), and copper and zinc naphthenate. Penta is most often used for posts set in fresh water. Posts treated with it take on a brown color. Penta slowly breaks down into biodegradable compounds, but meanwhile leaches into the soil and contaminates ground water, so it isn't likely to be used for this purpose much longer. Don't use penta-treated posts around livestock, especially animals that crib or lick.

Water-borne preservatives, also known by the scary-sounding phrase, "inorganic arsenical compounds," are the most widely used chemicals for fence posts, rails, and trellis stakes. They include chromated copper arsenate (CCA), ammoniacal copper arsenate (ACA), acid copper chromate (ACC), chromated zinc chloride (CZC), ammoniacal copper zinc arsenate (ACZA), and fluor chrome arsenate phenol (FCAP).

Of these, CCA is the most effective, longest lasting, and most economical. It is also considered the safest around plants and animals. Unlike creosote and penta, CCA won't chemically burn plants that touch treated trellis posts, provided they've been washed or slightly weathered. CCA is odorless, becomes insoluble after treatment, and does not leach out of wood. No one knows how long CCA-treated posts will last. Test stakes planted in the ground in the 1930s, although not entirely immune to boring insects, haven't rotted yet. Predictions are they could last one hundred years.

CCA usually leaves no residue on the surface of wood, which is why posts treated with it are considered safe to handle. A notable exception is a post with a white or gritty surface, which you should avoid buying. If you do get stuck with one, handle it with gloves, and wash off any residue that gets on your skin.

A new form of pressure treatment, soon to be available, consists of infusing freshly harvested posts with diffusible borate salts, currently sold under the trade name Tim-Bor. The product is registered with the Environmental Protection Agency as a nonrestricted pesticide, meaning it is considered sufficiently safe that you don't need a license to buy or use it.

Posts treated with borates won't last as long as CCA-treated posts, because borates have little effect on soil-borne soft-rot fungi. As a first line of defense against these fungi, and also to repel water from the outside and prevent leaching of the borates from within, treated posts may be coated with the over-the-counter wood preservative copper naphthenate.

It is a good idea to wear gloves when you handle any treated posts, and to wash your hands afterwards, especially before you eat. In addition, wash your work clothes separately, before putting them back on. If you have to do any sawing, do it outdoors and protect your face and eyes with a dust mask and goggles. Collect all sawdust and scraps and bury them in an approved landfill. Never dispose of pressure-treated wood by burning.

Be aware that some people react to wood preservatives no matter how careful they are. You can get additional details from consumer protection sheets, available at no cost wherever pressure-treated wood is sold.

Treated Posts

Dense woods like Douglas fir are difficult to pressure-treat. Douglas fir and other Western woods are usually treated with ACA or ACZA. Other woods, although not as dense as Douglas fir and therefore not as strong, are easy to treat. Southern yellow pine, usually treated with CCA, is the most treatable of all woods because its cellular structure allows the best penetration. Sapwood is less dense and therefore easier to treat than heartwood. Treated sapwood, incidentally, is more resistant to rot than untreated heartwood.

ACA and ACZA tint wood brown; CCA tints it a light blue-green. Either color can easily be covered with an oil-based, semi-transparent paint or stain after the wood has been allowed to weather for six months to a year. Coating treated posts and rails with paint, stain, or other water repellent keeps them from splitting due to swelling and shrinking during weather changes. Although water-repellent coatings do not themselves preserve wood, they help the wood retain a preservative. Some time soon, we'll be able to buy posts treated with preservative and water repellent at the same time.

Only a few species of wood are naturally more resistant to decay than pressure-treated varieties. Among them are black locust and osage orange. Imported South American hardwoods and Australian ironbark, a form of locust sold under the trade name Insultimber, make posts that last one hundred years or more. Untreated red cedar posts are also fairly durable. Although they generally don't last as long as pressure-treated posts and they're more expensive, they nonetheless remain popular because they don't bend easily.

Most treated posts are made of southern yellow pine, which tends to twist and warp over time. Also called "hard" pine, southern yellow pine is not a kind of tree but a whole group that includes longleaf, shortleaf, loblolly, and Virginia pines. Posts made from these species are less prone to breakage than those made from other pines like white or northern red, because they are tougher and their branches do not radiate from one central point as do those of the latter two species. Southern yellow pine does, however, split easily unless you take preventive measures, and moisture collecting in the splits leads to early decay.

Preservative-treated wood comes in different grades, ranging from lower grades suitable for field fences to higher grades offering greater aesthetic appeal. The amount of preservative used, and its degree of penetration, also vary. The American Wood Preservers Bureau has developed a marking program to help you identify the kind and level of treatment.

If you're buying milled posts, look for posts labeled "ground contact," "LP-22," or "FP" (for "fence post"), meaning that more chemicals have been forced deeper into the wood. Wood suitable for rails is labeled "above ground." These words appear on a quality mark stamped on the end of the post or on a tag stapled to it. Also look for "KDAT" on the label, meaning the posts were kiln dried after treatment.

Watch out for posts labeled "treated to refusal," which usually means the wood could not be adequately treated due to its species or its high moisture content. Properly treated posts are kiln dried before treatment as well as afterwards. If a post is treated while it is still green, chemicals cannot penetrate to the core. The center of the post will therefore slowly rot away until one fine day the post crumbles, and you'll be left wondering why.

Wood preservatives can, of course, be brushed, dipped, soaked, or sprayed on. But treating posts yourself can be a waste of time and money since the degree of protection is often shallow and unpredictable. The longest most hand-dipped posts will last is ten years. Moisture and bacteria remaining within the post will leave the core exposed to rapid decay. By contrast, pressure treatments usually penetrate the wood uniformly, establishing an environment that discourages fungi and insects.

The one possible exception is soaking freshly harvested, debarked posts in a solution of borate salts (Tim-Bor). Wood with a moisture content of at least 40 percent will pick up borates from the surface and diffuse them by osmosis to the center,

making the post less vulnerable to decay and insect damage. To further increase the post's rot resistance, butt-dip it in copper naphthenate. Soak the post overnight, coating it to a to a few inches above ground.

Other highly touted do-it-yourself methods worth mentioning, if only because they don't work, are coating the lower portion of posts with asphalt or tar and charring posts in hot ashes. By causing a reduction in diameter, charring may actually weaken a post.

Although pressure treatment increases the cost of posts, your ultimate consideration should be cost per year of use. The more expensive posts that last longer are actually cheaper. Any wooden part of your fence touching soil should be pressure treated. If you live where rain, fog, or high humidity are normal conditions, wood rails should be pressure treated as well.

Wood Post Fasteners

Fencing material can be fastened to wood posts with staples, nails, or bolts. Staples are used for wire fencing and plastic netting. Nails and bolts are used to attach wood to wood, such as for a rail fence. If you use untreated posts, always trim off sapwood before fastening any fencing material.

An alternative to using fasteners is to notch posts with slots to hold line wire or drill posts through, so wire or rails can be inserted without any fasteners. If you're putting up a wire fence, you should know that holes through posts will hold moisture that speeds wire corrosion.

Staples

For safety's sake, don't be tempted to walk around with your mouth full of staples. Don't put them in your pocket, either, or you'll whisk them back out the first time you bend down to pick up a tool. Carry staples in a metal can, in a plastic carrier, or in a carpenter's apron.

Staples come in two designs, U-shaped and L-shaped. L-staples won't work loose as readily, but aren't available in all areas. They come in lengths from 1¼ to 2 inches (30 to 50 mm). U-staples may have either smooth or barbed shanks. For softwood posts, barbed staples work best because they're more resistant to pull-out.

U-staples vary in length from ⅜ to 2½ inches (9.5 to 63.5 mm). Thinner staples made from 13-gauge wire are for lightweight jobs such as attaching chicken wire. The most common size is ¾ inch (19 mm). The largest staples, 2½ inches (63.5 mm) long and made of 6-gauge wire, are designed for attaching stock panels. Staples designed for field fencing are 9-gauge and 1¼ or 1½ inches (32 and 38 mm) long.

You may find only one size staple available in your area; if you need a different size, you will have to order by mail. Since you want the staples to hold as long as your posts last, staple length is important. Use 1¼-inch (32 mm) staples for untreated hardwood. Since pressure treatment has a lubricating effect that reduces a staple's holding ability, use at least 1½-inch (38 mm) staples in treated softwood. Use still longer staples for untreated softwood, to reduce the chance they'll work loose. (On the other hand, untreated softwood posts won't last long, anyway, making the holding power of staples in them somewhat irrelevant.) For high-tension fences, use 1¾-inch (45 mm) staples.

Zinc-coated staples, called "galvanized," last longer than nongalvanized staples and won't as readily encourage adjacent wire to rust. Staples can be either hot-dip galvanized or electroplated, the main difference being that hot dipping "anneals" the staples, or renders them less brittle. Staples with the thickest zinc coating, designated "Class 3," last the longest. They definitely should be used with posts treated with corrosion-causing water-borne preservatives.

FENCE STAPLE SIZES

Length			Approximate Number	
inches	mm	gauge	per lb	per kg
¾	19	14	480	1055
1	25	9	105	235
1⅛	29	9	95	210
1¼	32	9	85	190
1½	38	9	70	165
1¾	45	9	65	140
2	51	9	55	125
2½	64	6	25	55

As a rule of thumb, you will need ½ pound (.25 kg) of staples for each 100 feet (30 m) of woven wire or 400 feet (120 m) of barbed or smooth wire. The accompanying chart shows the approximate number of staples per pound (kilogram) for each staple size.

Driving Staples

A bit of trickery is involved in driving U-shaped staples. First, to avoid splitting the wood grain, alternate staples ½ inch (1 cm) to the right and left of an imaginary center line as you go down the post. Second, never drive in a staple vertically, with both legs in the same grain, or again you will split the wood and the staple will pop loose.

If you look closely at a staple, you will see that the ends are pointed and each has one flat face. Every staple is either left cut or right cut. To determine which, lay the staple in the palm of your hand with its legs pointing in the same direction as your fingers. Logically enough, if the cut face is on the left side, the staple is a left cut. If the cut face is on the right, it's a right cut.

Although few fencers are willing to take the time, sorting your staples lets you increase their holding power by 40 percent. Position each staple vertically along the post, then rotate the top leg 45° *away* from its cut face before driving it home. Those flat surfaces cause the legs to curve as you drive the staple in — each leg curving away from its flat face. If you rotate the staple the wrong way, the legs will curve toward each other and cross inside the wood, and the staple won't hold nearly as well. Ideally, right-face and left-face staples should be alternated as you work down the post.

Push the wire against the post with your arm, knee, or boot and then drive the staple in. If you use the staple to pull the wire to the post, the wire may get away from you and, like a slingshot, send the staple flying.

Set staples snugly at end posts, but not so tightly that you bury the staple in the post and damage the wire. Where wire changes direction, such as around a curve post, use two staples. Instead of driving the second one in, hang it loosely behind the wire to keep it from digging into the post, thereby reducing wear due to friction. If the wire goes around a full 90° corner, use two pairs of staples, side by side.

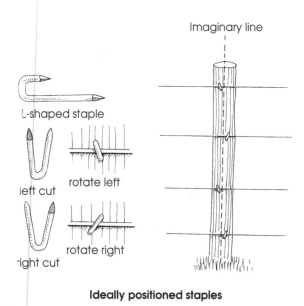

Ideally positioned staples

In line posts, set staples only so far that line wire can still slide freely. This allows room for the wire to move as the weather changes or as animals bump into the fence. Setting staples loosely also lets you periodically tighten a fence at the anchor post, instead of having to tighten each section between line posts.

Wire stapled to any post on a rise or knoll tends to pull downward. Conversely, wire stapled to any post in a depression or dip tends to pull upward. To prevent this stress from pulling the staple out, add a second, horizontally oriented staple where you want the fence wire to meet the post. For a rise, rest the fence wire on the horizontal staple and fasten it in place with a second staple. For a dip, place the fence wire *beneath* the horizontal staple and fasten it in place.

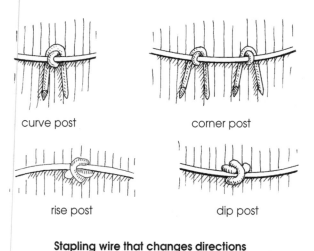

Stapling wire that changes directions

Nails

"Common" nails used for rail fence construction are thicker than staples; therefore they're more likely to split a post. Splitting reduces holding power and causes moisture to collect, leading to more rapid decay. To reduce the chance of splitting a round post, "face" it by shaving off a slab to provide a flat nailing surface. Other ways to reduce the chance of splitting are to stagger the nails and drill pilot holes or blunt the pointed end of each nail by placing the head against something solid and hitting the point once with a hammer.

Although blunting decreases the chance of splitting, it also decreases the nail's holding power. For better holding power, use fluted shank, spiral shank, or ring shank nails instead of smooth shank common nails. To avoid unsightly rust streaks, buy aluminum rather than steel nails. Even galvanized steel nails will eventually rust. Aluminum nails are softer and more expensive per pound than galvanized ones, but since they are lighter, you get more of them so it's pretty much a wash.

Bolts

As an alternative to nails, you might use bolts. They are much more expensive, but they let you readily remove and replace damaged rails. Two kinds of bolt are used for rail fence construction: lag bolts and through bolts. Both come plain or electroplated, the latter being slightly more resistant to rust.

Lag bolts look like oversized screws and, in fact, are sometimes erroneously called "lag screws." In addition to being used in rail fences, they are also used in wire fences to fasten wood braces to anchor posts.

A through bolt, as its name implies, goes all the way through the post and is held in place by a nut threaded onto the end. Fencers use one of two styles: carriage or machine. Both require drilled holes.

A carriage bolt is slipped into the hole and then hit with a hammer until it bites into the wood and won't turn when you thread a nut on the other end. A machine bolt has a square or hexagonal head you grip with a wrench so it won't turn while you thread a nut on the other end. As a cheaper alternative to through bolts, use threaded steel rods, cut to length

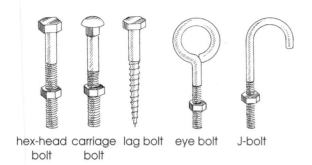

hex-head bolt | carriage bolt | lag bolt | eye bolt | J-bolt

with a hacksaw and secured with a nut and washer at each end.

To better distribute stress, place a washer under the head of a lag or square-head bolt, and one at the end of any through bolt, before threading on the nut. Tighten bolts just until they are snug, then go back and retighten a month or so later, after the wood has had time to shrink in the sun.

Two other kinds of bolt are used with wood posts: J-bolts and eye bolts. J-bolts have one end shaped into an open loop. They are used to build stock panel corrals. Eye bolts have one end shaped into a closed loop, and are used in cable corrals.

Steel Posts

Steel posts are less expensive than wood posts. They don't require dug holes, as wood posts sometimes do, but instead are driven into the ground. Although they will rust after the paint wears off, they won't rot away like a wood post will. Contrary to what you might expect, steel posts aren't nearly as strong as wood posts. It takes 225 pounds (49.5 N) of steady force to break the smallest wood post of 2½-inch (64 mm) diameter, while a standard T-post will give under only 63 pounds (14 N).

Steel posts can be used as anchor posts as well as line posts for low-tension wire fences, but are more often used as line posts in conjunction with wood anchor posts, especially where the soil is sandy or moist, where animals are closely confined, and for wire fences under high tension. By adding brackets designed for the purpose, they can also become line posts for rail fences. For use around horses, steel posts should be fitted with plastic caps to prevent injury.

Steel posts are made in several different shapes and sizes. T-posts, so-called because in cross-section they are shaped like the letter T, are the most common kind of post for field fences. They generally come in lengths ranging from 5 to 8 feet and every 6 inches in between, although you can get posts as long as 9 or 10 feet. The most popular lengths are 6½ and 7 feet (about 2 m). In countries using metric, you are likely to find steel posts running in 15 cm intervals, from 135 to 240 cm high. Widths vary from about 1¼ to 1¾ inches (32 to 44 mm). You don't always know what width you are getting, which can create problems if you're building an electric fence with snap-on insulators (described in chapter 9).

In some areas you can also find forged U-posts, angle-iron posts (which, for consistency's sake, you'd think would be called "L-posts"), or Y-posts. The latter, sometimes called "star" posts, are relatively scarce in the U.S., which is surprising because they are strong enough to be used as gate posts and as anchor posts for a high-tension fence.

Angle-iron posts, like T-posts, come in a narrow range of sizes. U-posts, on the other hand, are of three distinct sizes: heavyweight for field fencing

U-POSTS

Weight	Gauge	Width	Thickness	Height
Heavy	12	2¹⁄₁₆ in. 52 mm	1 in. 25 mm	5 ft. to 8 ft. 150 to 245 cm
Medium	13	1¾ in. 45 mm	⅞ in. 22 mm	5 ft. to 8 ft. 150 to 245 mm
Light	14	1¼ in. 32 mm	⁹⁄₁₆ in. 14 mm	3 ft. to 6 ft. 915 to 1830 mm

and other situations where strength and longevity are important; mediumweight for temporary fences and for use with woven wire; and lightweight for electric fences, garden fences, and for use with lightweight netting.

Most steel posts have an arrow-shaped anchor plate bolted, clamped, or riveted near the base. This plate increases the post's contact with the soil, helping to keep it from being pulled out. When you drive the post in, it will tend to turn unless you brace a booted foot against the anchor plate. The anchor plate should go 2 to 3 inches (50 to 76 mm) below ground.

U-posts and angle-iron posts have little tabs, called "lugs," along their face. Hook the line wire in, then close the lug by tapping it with a hammer. A T-post has bumps, called "studs," jutting from its face. When you set steel posts into the ground, be sure all the lugged or studded faces are on the wire side of

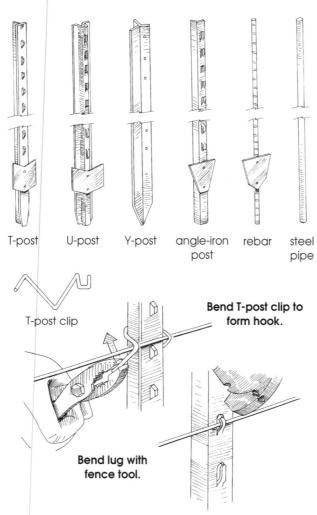

T-post   U-post   Y-post   angle-iron   rebar   steel
post   post   pipe

T-post clip

Bend T-post clip to form hook.

Bend lug with fence tool.

the fence — facing toward the inside if you want to keep livestock in, toward the outside if you want to keep animals out, as for a garden.

The lugs of a T-post help you accurately position fencing wire, which is held in place with wire T-post clips. Hook a clip over the line wire, snap it around the post, and use pliers to bend the other end around the wire to hold it in place. If you use Class 3 galvanized clips of the same size as your fence wire, the clips will share the same life expectancy as

the rest of your fence. If, however, you force clips into the wire, bending it, the wire's galvanizing may crack and your fence will rust more easily.

For an electric fence, you'll need insulators instead of wire clips (see chapter 9). Although insulators add to the cost of a fence, they are often used instead of clips, even for nonelectric fences. Insulators hold fence wire away from posts, while clips allow the wire to rub against posts, causing premature rusting.

Steel rods sold to reinforce concrete may be cut to length and used for an electric fence. Most farm stores carry ready-made electric fence posts consisting of 48-inch (120 cm) lengths of 3/8-inch (9.5 mm) rebar with anchor plates attached. Sometimes the anchor plates are welded on; sometimes they are simply slipped over the rebar, so that as you pound the posts into the ground, the anchor plates uselessly slide upward. Fence posts made with rebar aren't very sturdy and are usually designed for short-term use.

Ordinary steel pipe can be used to construct a sturdy fence, and an economical one if you have access to an inexpensive source of used pipe. You will need at least 1¾-inch-diameter (45 mm) pipe for line posts, larger pipe for anchor posts. For heavy-duty anchor posts, fill 6- to 8-inch-diameter (15 to 20 cm) pipe with concrete. Pipe posts are best suited for woven wire or rail fences.

Stone Posts

Posts made of limestone were once quite common in Kansas. Cut granite is still sometimes available in northern New England, Quebec, and the Maritime Provinces. Stone posts have the distinct advantage of not rusting like steel and not rotting, burning, bending, or splintering like wood. They will, however, chip, crack, or shatter if you bump into them too hard, and they are very expensive. They are primarily reserved for formal rail fences on level terrain. You may find them at larger garden centers, usually with slots carved out for easy rail positioning.

Rock Corner Posts

Unlike a stone post cut from a solid block, a rock corner post consists of loose rocks tossed into a

A rock corner post

cylinder of woven wire. The cylinder, about 3 feet (1 m) in diameter, is firmly affixed to a solid wood corner post. This structure serves two purposes — it provides a place to throw large rocks and it increases the stability of the corner post.

Concrete Posts

American farmers were using concrete posts as early as 1910. Sears Roebuck soon got into the act by offering reasonably priced do-it-yourself molds. Because concrete posts are strong and durable, and they don't use up a scarce natural resource, they are once again becoming popular. They are used primarily for trellises, rail fences, and field fences.

Concrete posts offer all the advantages of stone posts, plus a few additional ones. They are more readily available (especially if you make your own), they are cheaper, and they don't chip or crack as easily. They are especially suitable as anchor posts and as line posts on a curve, where wire tends to pull other kinds of post out of line.

Shapes vary according to the type of fence under construction. For a rail fence, posts may be slotted for each rail or they may be H-shaped, with rails slipped into the channels and separated by spacers. For a wire fence or trellis, posts may be notched along one face or have holes molded in them for attaching wire clips.

Like cut stone posts, concrete posts are available at well-stocked garden stores. You might purchase them directly from a manufacturer, if one is nearby. A single post can weigh as much as 300 pounds, so freight can become costly for posts transported any significant distance. If you make your own, you will have straight, sturdy, long-lasting posts at an economical price.

Plastic Posts

Plastic posts are designed for two distinct purposes: rail fences and temporary electric fences. Posts for rail fence systems are described in chapter 12, "Rail Fences." Plastic posts for electric fences are relatively inexpensive but don't have much physical strength. They bend and break quite easily, and have a life span of only three or four years. Since plastic is self-insulating, you don't need additional insulators.

A plastic tread-in post is molded with a steel spike at the bottom and a foot pad you step on to push the post into the ground. The post easily slips into loose soil or loam, but is difficult to use in hard-packed clay. Each post has molded holders for wire and a top loop that lets you use one post to brace another at corners and ends.

As an inexpensive alternative to commercial plastic posts, make plastic O-posts from 5-foot (150 cm) lengths of 1-inch-diameter (25 mm) PVC water pipe. These posts can be used only seasonally

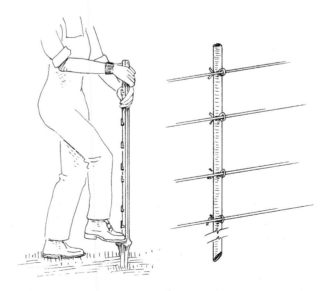

Left: plastic tread-in post. Right: plastic O-post.

where freezing weather is likely. Otherwise, water accumulating in the hollow center could freeze, shattering the posts.

If you haven't any way to transport the 20-foot (6 m) lengths that PVC pipe comes in, bring along a measuring tape and a hacksaw with a fine-toothed blade and cut the pipes to length at the building supply center. Cutting the bottoms at a 45° angle will help ease the posts into the soil.

Drill holes through the pipe at intervals that are appropriately spaced for your fence design. To attach line wire, loop a 6-inch (15 cm) length of smooth wire around the line wire, thread the ends through a hole in the pipe, and bend them back in the fashion of a cotter pin. Set posts no farther apart than 10 feet (3 m).

Fiberglass Posts

Fiberglass posts are very much like plastic posts in being lightweight, inexpensive, self-insulating, and used primarily for electric fences. They are very much *unlike* plastic posts in being sturdy, flexible, and long lasting. If you shop knowledgeably, you can get fiberglass posts that should last virtually forever.

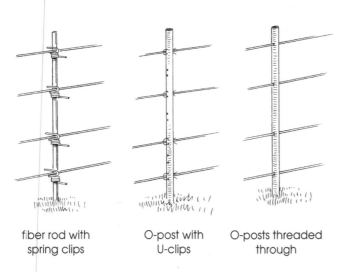

fiber rod with spring clips   O-post with U-clips   O-posts threaded through

The first fiberglass fence posts, appearing in 1973, were T-shaped, in imitation of steel T-posts. Soon the fencing market was flooded with posts made from discarded fiberglass "sucker rods," designed as pistons for oil well suction pumps. The greater flexibility of these solid-core round posts led to the development of larger, hollow O-shaped posts in 1983.

Today, the most common fiberglass posts are round and either solid or hollow at the core. They come in three distinct grades. The lowest grade consists of fatigued, overstressed sucker rods sold to unwary fence builders, giving new meaning to their original name. Typically, these solid posts are greenish gray, ⅞ to 1¼ inch (22 to 31 mm) in diameter, and cheap. Since they are not designed to withstand sun exposure, they begin to "blossom" with fibers sticking out all over after only three to six months. Handle a blossoming post and you will get nasty slivers in your hand. The protruding fibers collect dust and moisture, the post loses its insulation properties, and your electric fence gets shorted out. On the other hand, if you are building a permanent electric fence on open range, where posts don't require frequent handling, you can use sucker rods to cut costs and still have an effective fence.

The most common grade of fiberglass posts are "ultraviolet inhibited." These posts are molded with an inhibitor mixed into the resin to retard deterioration as a result of sun exposure. The key word here is "retard."

The third grade is "ultraviolet coated." These posts are coated with a thick layer of polyurethane that supposedly won't break down at all; theoretically these posts will last forever. Some experts predict they will last longer than wood or steel. Others say they have seen the coating flake off. Some manufacturers back coated posts with a twenty-year warranty.

The best way to know if you are getting top grade posts is to ask for a written warranty. To otherwise determine if you are getting a quality post, run your hand along it. The post should feel smooth, with no bumps or prickly fibers sticking out. You should also be able to bend the post quite far without having it splinter or shatter.

Fiberglass Designs

Solid fiberglass rods, sometimes called "fiber rods" or "rod posts," are ¼ to 2 inches (6 to 50 mm) in diameter or less. The thinnest ones are usually pencil-pointed, making them easy to push into soft soil. They don't have much holding power, though, so if you use them in loose or sandy soil, get long ones and drive them in deeply. Alternatively, you can increase a post's stability by fitting the bottom with a spring steel foot, available from suppliers of fiberglass posts.

Some posts have holes drilled in them. You can either thread line wire through the holes or attach it with cotter pins. For greater flexibility in wire heights, use adjustable torsion springs, called "spring clips."

These stiff wire coils slip over the post and grip it tightly. The ends curve back to form two loops that you have to squeeze together to release the clip from the post. The loops are big enough to thread line wire through and let it move freely. Better brands won't slide up and down unless you release the spring; some brands are so tight you need pliers to move them.

Fiber rod posts range in length from 4 to 12 feet (120 to 360 cm). They are handy for all sorts of things besides fencing. You can use them to stake tomatoes, peppers, or newly planted fruit trees. You can set them out as a visual aid to mark off newly seeded lawn. You can slip a short piece of garden hose on one end as a grip to make a dandy walking stick for jaunts in the woods.

A solid-core rod is more flexible than a hollow O-post and has a better "memory." If a heavy animal galloping at full speed knocks one down, the post will pop back into place as soon as the animal has passed. Even if a post is flattened by a fallen tree or bent under the weight of accumulated ice, it will pop back up when the stress is removed.

O-posts, on the other hand, are easier to see because they come in larger diameters — 1 inch and 2 inches (25 mm and 50 mm). They are also longer, ranging from 8 to 10 feet (240 to 300 mm). They are designed for permanent electric fence systems. Some have holes drilled in them so you can either thread line wire through or attach it with U-clips fastened through pairs of holes. Since fiberglass posts don't have much holding power, it's a good idea in loose or sandy soil to add anchor plates (available from O-post suppliers); or set the posts in concrete if you use them in anchor positions.

Fiberglass T-posts, like metal T-posts, have studs molded in them to hold line wire in place. These posts aren't popular because they tend to split. Furthermore, their rigid clips don't allow line wire to move freely, creating problems both when the wire is initially strung and later when it stretches and shrinks in changing weather.

POST SELECTION GUIDE

	Wood	Steel	Concrete	Plastic	Fiberglass
MOSTLY USED AS					
Anchor Posts	X		X		
Line Posts	X	X	X	X	X
Spacers	X			X	X
MOSTLY USED FOR					
Barbed Wire	X	X			
HT Smooth (N)	X		X		X
HT Smooth (E)	X				X
LT Smooth (E)	X	X	X	X	X
Woven Wire	X	X	X		X
Plastic Net (N)	X	X			X
Plastic Net (E)				X	X
Rail (N)	X	X	X	X	X
Rail (E)					X

LEGEND:

HT – high-tension   N – nonelectric
LT – low-tension   E – electric

Spacers

Lightweight spacers, also called "stays," "battens," or "droppers," are used in wire-strand fences to keep line wires properly spaced, to distribute impact among all the wires, and to increase the fence's visibility. Using spacers lets you place line posts farther apart, thereby reducing your overall cost.

Spacers are set shallowly into the ground for a standard fence but don't touch the ground in a suspension fence. Spacers that don't touch the ground are sometimes called "dancers," because they bounce up and down whenever the wires move.

Dancers are firmly attached to the line wires and, on impact, move along with them. By contrast, a line post (or a spacer that touches the ground) is stationary so that moving line wires slide past it. Unlike line posts, which generally go on the side of the fence experiencing the most animal pressure, dancers are attached on alternate sides for increased strength.

Spacers are added last, after the line wires have been strung and pulled tight. They come in various lengths to suit fences of various heights. They may be made of high-density hardwood, fiberglass, plastic, or heavy-gauge wire.

Wood spacers may have holes drilled in them so the line wires can be threaded through, or they may have grooves notched in them. If the groove is angled, you don't need additional clips to hold the line wire in place. Straight-grooved spacers require clips. Both kinds of notched spacer are firmly affixed to line wires and move along with them as the fence expands and contracts.

Fiberglass spacers also come notched. More commonly, lightweight fiber rods are used, pushed 8 to 12 inches (20 to 30 cm) into the ground, with spring clips that let the line wire flow freely. Plastic posts, homemade or commercial, can also be used as spacers. You can fabricate spacers of PVC water pipe by sawing grooves for the line wires to rest in and using clips to hold the line wire in place. Because of their weight, these spacers will touch the ground.

Commercially made wire spacers consist of spiraled 9½-gauge wire twisted together into 38-, 42-, and 48-inch (90, 105, and 120 cm) lengths. You can make spacers from any heavy-gauge wire by carefully looping it around each line wire to preserve proper spacing.

Wire and wood spacers are best reserved for nonelectric fences. They will work for an electric fence *only* if the spacers don't touch the ground *and* all the line wires are electrified. Since fiberglass and plastic spacers are self-insulating, they can be used for electric as well as nonelectric fences.

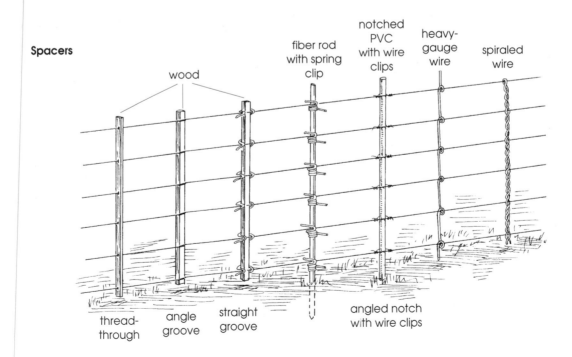

Spacers

wood — fiber rod with spring clip — notched PVC with wire clips — heavy-gauge wire — spiraled wire

thread-through — angle groove — straight groove — angled notch with wire clips

CHAPTER 3 ▾▴▾▴▾▴▾▴▾▴▾

LAYING THE FOUNDATION

▾▴▾▴▾▴▾▴▾▴▾▴▾▴▾▴▾▴▾▴▾▴▾▴▾

No matter what kind of permanent fence you put in, you will set your anchor posts first, then space your line posts between them. Establishing where to put anchor posts is fairly easy, since their positions depend almost entirely on such physical features as where corners and gates are, how far the fence runs between corners and gates, and whether the terrain is level or hilly. Anchor posts, then, form your fence's foundation.

ANCHOR POSTS

Anchor posts include all end, gate, and corner posts and any post that falls in a sharp curve, dip, or rise. An end post, as the name implies, occurs at the end of a fence run, where fence materials pull from one direction. The end post might be located next to a building or another, perpendicular fence, or at either end of a trellis.

A gate post is a special kind of end post that is not just pulled by the fence but also has to support a swinging gate. Unless you use a double gate, technically only the post on the hinge side is a gate post; the post on the latch side is an end post.

A corner post is any post located where the fence changes direction, pulling the post two ways. If the direction change is greater than 5°, the corner post must either be set in concrete or braced, thus becoming an anchor post. A curve post is essentially one in a series of corner posts occurring where a fence changes direction gradually rather than abruptly.

Dip and rise posts are located in low and high points on hilly terrain. For fence wire strung under tension, dip posts must be anchored so upward pull doesn't yank them out of the ground, and rise posts must be anchored so downward pressure doesn't pull them out of plumb.

Anchor posts differ from line posts in four ways: they're stouter; they're longer and set deeper; they're set more firmly (often with concrete footings); and they're usually braced.

Brace Assemblies

How much bracing each anchor post needs depends on the kind of fence you are constructing. For a standard rail fence, gate posts are the only ones that require bracing, since the fence gets most of its rigidity from the structural nature of the rails. Posts in the remaining anchor positions need only be set in concrete (see chapter 4 for details).

A wire fence, by contrast, has to be strung taut to function properly, which puts stress on *all* the anchor posts. The anchor posts of a wire strand fence must therefore be able to withstand the pressure exerted by wire under tension, multiplied by the number of horizontal wires in the fence. How much tension this adds up to depends on whether the wire is under low tension or high tension. A low-tension fence is stretched just enough to prevent sag. A high-tension fence is stretched tight enough to bounce back on impact.

Anchor posts must be able to withstand additional stress imposed by the contraction of wire in cold weather and by impact from animals and sometimes vehicles or equipment. The more line wires you have and the more tension your fence is under, the stronger your brace assemblies must be to prevent your anchor posts from leaning or being pulled out of the ground. A fence in sandy or swampy soil requires even stronger bracing than one in firm soil.

The four basic kinds of brace assembly are

deadman, diagonal stay, diagonal brace assembly, and horizontal brace assembly. Tie-back systems, wherein posts are anchored with guy wires (utility poles are one example of this), are not included here because they waste space and make weed control difficult. In addition, because the guy wires are hard to see, they easily get bumped into and damaged by mowers or tripped over by unwary workers.

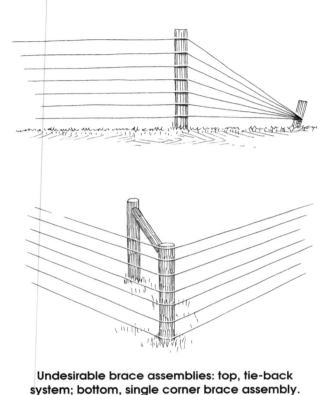

Undesirable brace assemblies: top, tie-back system; bottom, single corner brace assembly.

Similarly, using a single brace assembly at a corner post creates more problems than it solves. This method is sometimes touted as a way to reduce cost and time. But, because it blocks the inside of the corner, the brace virtually ensures that one day the corner will become a thicket of weeds and brambles. The only sensible bracing system is one that lies flush with the fence line.

Deadman

A deadman, sometimes called a "bedlog," consists of a short length of pressure-treated lumber nailed toward the bottom of a wood post to form a T. It works on the same principle as the anchor plate at the bottom of a T-post, and is used in low-tension fences and trellises to add stability to hand-set posts and those in sandy soil.

Theoretically, the strongest deadman consists of two 4-foot-long (120 cm) 2x6s (5x15 cm) for each post. With galvanized nails driven at an angle, attach one cross piece 6 inches (15 cm) from the bottom of the post. Nail the second piece parallel to the first, 12 inches (30 cm) up and on the opposite side (with the post between them). Plant the post in a 2-foot-deep (60 cm) trench, or deeper if necessary to resist frost heave. Orient the cross pieces perpendicular to the line of pull, the upper one facing the direction of pull exerted by the line wire.

This double deadman isn't as strong in reality as it is in theory because, in order to work properly, the deadman must be positioned against firm soil. In digging a trench sufficiently large for a double deadman, you are bound to disturb too much soil, and the resulting backfill weakens your brace.

A better alternative is a single deadman consisting of a 4-foot (120 cm) length of pressure-treated

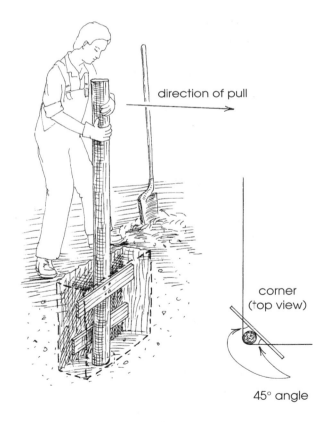

direction of pull

corner
(top view)

45° angle

Deadman

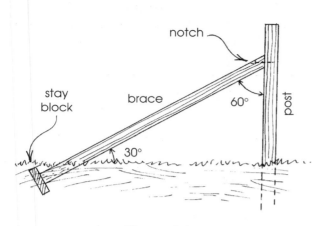

Diagonal stay

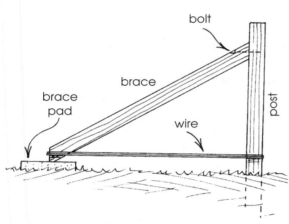

Floating brace

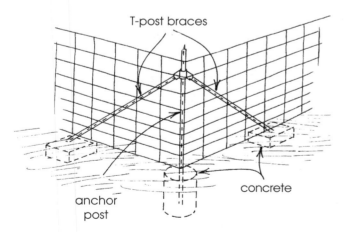

T-post stays

4x4 or 2x8 (10x10 or 4x20 cm) nailed to the post where it will fall just below soil surface. After digging the post hole, drop the post in, draw an outline where the cross piece falls, and dig a trench only long enough and deep enough to accommodate the deadman. Position it against solid soil, facing the direction of pull. If you are bracing a corner post, position the deadman so it faces half-way between the two directions of pull.

Diagonal Stay

A good deal stronger than a deadman is a diagonal stay, also called a "one-post brace" or a "post and stay brace." It consists of a wooden pole or length of lumber wedged against the top of the anchor post at one end and jammed against a stay block at the other. The brace should follow the fence line and should be on the *inside* of the fence, unless you're confining animals with a penchant for climbing.

The brace should be at least 4 inches (10 cm) in diameter and at least 2 feet (60 cm) longer than the part of the anchor post above the ground. Trim the upper end so the brace fits against the post at an angle of about 60° and secure it with a ⅜-inch-diameter (1 cm) lag bolt or a long galvanized nail.

To keep the brace firm, notch the post slightly where it meets the brace, which should be about halfway between ground level and the top of the fencing material. Much higher, and the brace will put too much pressure on the post; much lower, and it will put too much pressure on the stay block.

At the ground end, set the brace in a shallow V-shaped hole against a stay block consisting of a flat stone, a solid concrete block, or a block of pressure-treated wood butted against undisturbed soil. The brace should fit so tight that you literally have to shoe-horn it into place. If you are using diagonal stays to firm up a corner post, you will need two assemblies at right angles.

For a low-tension fence with steel T-posts, use a second T-post as your diagonal stay, attached to the anchor post with a T-post collar or bracket (sources are listed in the appendix). With a hammer, knock the anchor plate off the brace post.

To keep the assembly firm, set both the anchor post and the brace base in concrete. Where the brace touches the ground, dig a 15-inch-square (40 cm) hole, 18 inches (45 cm) deep, positioned so the brace enters the concrete 6 inches (15 cm) below ground and extends 6 inches (15 cm) into the concrete. If frost heave is a recurring problem, dig the hole deep enough so at least 8 inches (20 cm) of concrete extend below the frost line. Where the frost line is deep and the soil is rocky, this brace system won't be your best option.

A diagonal stay variation that doesn't require digging and that is suitable for high-tension fences is called a "floating" brace. Here, the brace rests on a pad consisting of either a pressure-treated 4x12 (10x30 cm), 2 feet (60 cm) long, or a reinforced concrete block measuring 4x8x18 inches (10x20x45 cm). The top of the pad must be even with ground level at the anchor post, requiring adjustment on a slope. The brace consists of an 8-foot-long (240 cm) pressure-treated 4x4 (10x10 cm), trimmed at one end to fit against the anchor post and bolted to it about 30 inches (75 cm) up, so it makes a 60° angle with the post and a 30° angle with the ground.

Tightly wrap two full loops of smooth wire around the post and the end of the brace, no more than 7 inches (18 cm) up from the ground. To keep the brace wire from having to bend around a corner, slightly round off the edges of the brace where the wire goes around it. Tighten the wire until the top of the post leans outward, using a wire stretcher or an in-line tensioner (described in chapters 5 and 6, respectively). Staple the brace wire to the post and the brace, but do *not* fix the brace to the pad.

The essence of a floating brace is that it is able to "float" along the pad's surface as wire tension changes in the fence. This brace requires less time, materials, and labor than a comparably strong H-brace and is best suited for shallow or loose soils, for hand-set anchor posts, and for high-tension fences with no more than eight line wires.

Diagonal Brace Assembly

A diagonal brace assembly, sometimes called a "two-post brace," is similar to a diagonal stay except that the lower end of the brace is wedged against a fence post for added strength. The second post, called a "brace post," can be smaller in diameter than the anchor post but should be larger than the line posts. For greatest strength, the brace post should be 8 feet (240 cm) from the anchor post. If stronger bracing is needed, add a second brace post 8 feet (240 cm) from the first to create a double diagonal brace assembly. Instead of measuring from the center of one post to the center of another as you normally would, measure from inside the anchor post to inside the brace post to avoid the extra work of trimming down the brace.

At the corners, you will need two diagonal assemblies at right angles to each other, following the fence line. If you are confining goats, always stretch fence material on the *inside* of a diagonal brace. Otherwise, the brace will offer the goats a ready-made way to climb up and out.

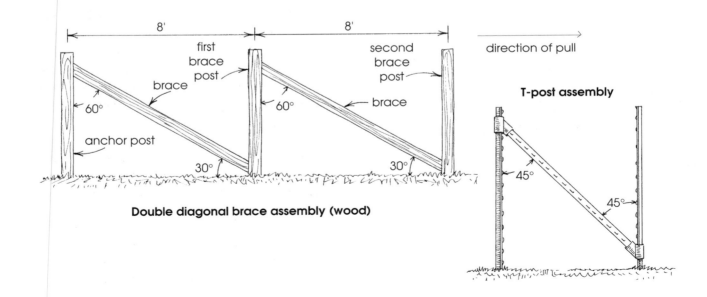

Double diagonal brace assembly (wood)

T-post assembly

Horizontal Brace

A horizontal brace, also called an "H-brace," "boxed assembly," "post and rail assembly," or "compression brace," is the strongest of all brace assemblies and therefore is best for high-tension fences. In moderately well-drained soil, a single H-brace assembly will support up to five strands of wire as a corner post and up to ten strands as an end post. Its strength comes from a wood or steel brace rail, or span, riding between the anchor post and brace post. The fence wire pulls against the anchor post from one direction while the rail pushes against it from the other.

The anchor post is usually 6 inches (15 cm) in diameter, the brace post 5 inches (13 cm). For the rail you will need a milled 4x4 (10x10 cm) or a 5-inch-diameter (13 cm) wood pole, at least 8 feet (240 cm) long. You can increase the holding capacity of the assembly by increasing the length of the rail to 10 feet (3 m). You will also need a pair of steel or fiberglass dowels or "brace pins," ⅜ to ½ inch (10 to 12 mm) in diameter, one 5 inches (13 cm) long and one 10 inches (225 cm) long. To save money, you can cut your own brace pins from ⅜-inch (10 mm) rebar.

Drill a hole in the anchor post in the direction of the fence line, 4 inches (10 cm) from the top, 2 inches (5 cm) deep, and the same diameter as your pins. Drill a hole of the same diameter and height, all the way through the first brace post. Bore holes of the same diameter, 2 inches (5 cm) deep into the ends of the brace. Drive the short pin into the anchor post and fit the brace into the protruding end. Lift the brace into place and drive the long pin into it through the brace post, leaving 2 inches (5 cm) of the pin sticking out the other side of the post.

To minimize lateral movement and keep the assembly rigid, you will need a brace wire, as well. The conventional way to add one is to wrap a length of 9-gauge galvanized smooth soft wire twice around the posts diagonally from the top of the brace post to the bottom of the anchor post on the side of the posts opposite the fencing material.

First, drive a horizontal staple on the back side of the anchor post, 4 inches (10 cm) from the bottom, sticking out far enough to keep the brace wire from sliding up the post. Wrap the smooth wire around the pin protruding from the brace post and staple it

down. Run the wire diagonally to the bottom of the anchor post and wrap it twice around the post, below the staple, run it back up to the top of the brace post. Wrap the end of the wire around the pin, and staple it down.

Tighten or "twitch" the wire by twisting it with a steel rod or a pressure-treated wood stick, about 24 inches (60 cm) long. The stick used for this purpose is called a "twitch" or "twist" stick. Facing the brace wire from the side of the assembly opposite the fence wire side, insert 2 inches (5 cm) of the stick between the two pairs of diagonal wires, about one-

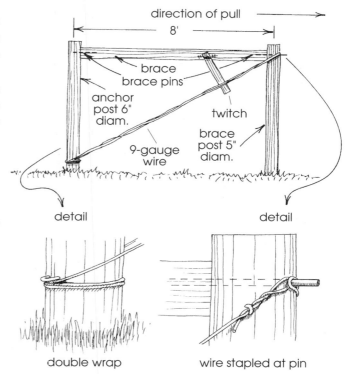

Horizontal brace

third of the way down. Using both hands, pull the twitch toward you to twist the diagonal wires together. Make five or six complete turns, or until the wire is taut. Take care not to overtension the brace wire or you could lift the anchor post out of the ground. Let the twitch come to rest against the rail so it can't unwind. To hold it in place, cut a short piece of smooth wire and staple it across the stick against the rail. Leave the stick there so you can adjust the tension as necessary.

A second method for installing brace wire, preferred by builders of high-tension fences, is to use a single strand of 12½-gauge high-tensile wire, strung diagonally from the lower end of the anchor post to the upper end of the brace post and tightened with an in-line tensioner (described in chapter 5). Although this method is more expensive, it is safer than twitching (the stick *can* get away from you) and it simplifies adjusting brace wire tension after the fence is up.

Adding a second brace wire along the opposite diagonal to form an X is, in most cases, unnecessary. It is also undesirable because it pulls the anchor post from the wrong direction. One exception is if you are bracing a gate post that supports an especially heavy gate, in which case add the second diagonal but tighten it only enough to hold the gate. Another exception is when you are bracing a pull-post (discussed below).

You can fashion an H-brace assembly for a low-tension fence, using T-posts, T-post collars or brackets, and a rail made from a T-post with the anchor plate knocked off. In this case, the distance between the anchor post and the brace post must equal the length of the brace.

Double H-Brace

Where the distance to the next anchor post is 165 feet (50 m) or less, one H-brace assembly at each end is plenty. If the distance is from 165 to 660 feet (50 to 200 m) or your soil is sandy and therefore has

MATERIAL LIST FOR H-BRACE ASSEMBLY

	Single Span	Double Span	Triple Span
8' 6" anchor post	1	1	1
8' 5" brace post	1	2	3
8' 4" brace pole	1	2	3
brace pin, 5"	1	1	1
brace pin, 10"	1	2	3
twitch stick	1	2	3
9-gauge HT wire	40' (12 m)	80' (24 m)	120' (36 m)
fence staples	5	10	15

For corner posts and pull posts you will need an assembly on each side.

less than average holding power, you'll need double assemblies to distribute the additional pull. You'll need double assemblies, too, at the corner posts of a high-tension fence with seven strands or more.

If the distance from one anchor post to the next is more than 660 feet (200 m), brace the posts with triple assemblies (or divide the fence with a pull-post assembly, described below). A multiple H-brace assembly consists of two or more single assemblies in a row. To keep the assemblies from buckling, make sure the anchor post and brace posts all lie in a straight line.

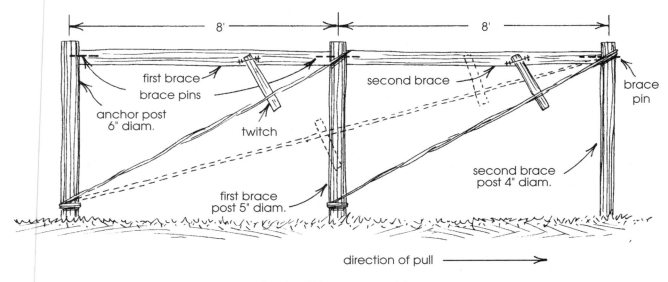

Double H-brace assembly

If your soil is light or moist, or wind tends to bank snow, leaves, or tumbleweeds against the fence, your anchor posts will experience extra stress. Increase the strength of a double assembly by extending the first brace wire to the second brace post. For a double-span assembly you would therefore need only one brace wire; for a triple assembly you would use two. Because of the intervening post, you will have to twitch the wire in two places. Staple the additional twitch stick to the first brace post, as shown in the illustration.

Pull Posts

In order to maintain proper tension in a long, straight run of wire fence, it is a good idea to add braced line posts periodically; these are called "pull" or "stretcher" posts. If you are using wood posts, the one that is braced should be at least 5 inches (13 cm) in diameter and the brace posts on each side of it should be 4 inches (10 cm). Bracing consists of two H-assemblies, each with a second diagonal stay wire added to form an "X" with the first diagonal wire, thus compensating for the pull of fencing from opposite directions.

The farther apart pull posts are spaced, the stronger the brace assemblies must be. If there is an anchor assembly within 250 feet (75 m) on both sides of the pull post, flank the post with single H-brace assemblies. If the distance is 250 to 500 feet

(75 to 150 m), use double assemblies.

When you brace a pull post with double assemblies, make sure the next nearest anchor post is also braced with double assemblies so it won't get pulled out of the ground. The reverse is not true — a double anchor assembly will not overpower a single pull-post assembly because wire pulls the post from two directions.

Dips and Rises

If you are constructing a fence over rough terrain, you will need a post at the bottom of every significant depression or dip to eliminate gaps, and a post at the top of every significant knoll or rise to maintain adequate fence height. A line post will do unless you're building a tension fence and/or the rise or dip is severe, in which case you'll need an anchor post. Dip posts must be well anchored to prevent pull-out, rise posts to prevent leaning.

For a wire fence, dip posts can experience a tremendous amount of pressure. Just how much is determined by degree of slope, measured according to the ratio of "fall" to "run," fall being the vertical drop and run being the horizontal distance over which the land slopes. The shorter the run with respect to fall, the more upward pull a wire fence exerts on a downslope post. Upward pull increases if the fence goes back uphill on the other side of the dip post. Even on a slight slope, wire can

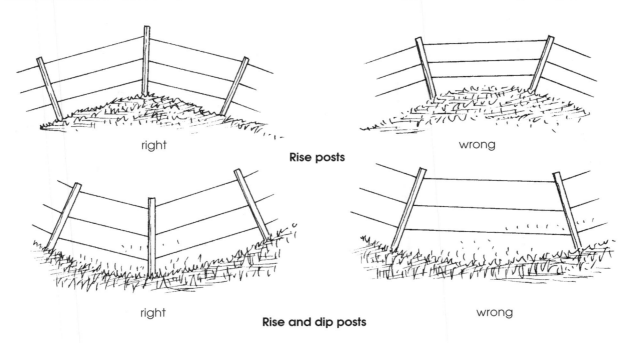

right   wrong
Rise posts

right   wrong
Rise and dip posts

exert tremendous pull on a downhill post. When tightening wire along a slope, always stretch it from the downhill direction. (Procedures for stretching wire are discussed in chapters 5 and 6.)

To anchor a single wooden rise or dip post, nail an 18-inch (45 cm) deadman 2 inches (5 cm) from

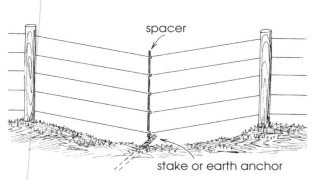

Stringing line wire across a shallow dip.

the bottom, oriented parallel to the fence line. If you are using concrete posts, insert an 18-inch (45 cm) length of rebar through a hole at the bottom. If you are building a rail fence with vinyl posts, drill a hole and insert an 18-inch (45 cm) steel rod or pipe. When setting dip posts, imbed the deadman in concrete.

For fences with fiberglass posts, anti-sink/tie-down disks are available that, used in combination with stakes or earth anchors, let you secure the

posts on rises and dips. If you are building a wire strand fence across a dip that's less than 2 feet (60 cm) deep and 6 feet (180 cm) across, use a spacer in place of a dip post and anchor it either with a metal stake driven into the ground at a 45° angle or with an earth anchor.

Earth Anchors

The primary use of earth anchors is to provide a place to attach the lower ends of guy wires. Look closely at a utility pole: wherever a guy meets the ground there is evidence of the earth anchor below. For fencing purposes, earth anchors are used mainly to hold down dip posts or to pull down line wire where they cross a depression.

To strengthen a dip post, place an earth anchor 2 to 3 feet (60 to 90 cm) down the fence line on both sides of the post and attach the anchors to the post with guy wires. To pull down line wires, fasten the wires to a spacer, place an earth anchor directly below the spacer, and attach the two together with a short length of wire.

Earth anchors are available in two styles. The conventional kind is called a "screw" anchor because you work it into the ground (or remove it) with a screwing action. It is topped with a shaft ending in a loop for fastening guy wire.

Screw anchors come in different diameters —

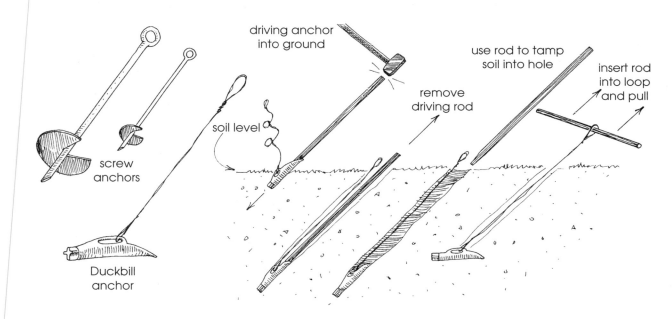

Earth anchors

the looser your soil, the larger you will need. The anchor should be driven into undisturbed soil to the depth of its shaft, which requires a power-driven auger.

A completely different design that's considerably more expensive but easier to install is sold under the trade name Duckbill. This anchor has a scoop on one end, shaped like a duck's bill, and is blunt on the other end. It has a little handle to which you must attach a length of galvanized steel cable or high-tensile wire. If you use high-tensile plastic wire, it will neither rust nor short out an electric fence.

To install a Duckbill anchor, insert a steel rod into the blunt end and drive it into the ground with a sledge hammer or post driver as far as it will go. Pull the rod out and use it to tamp soil into the hole. Make a loop at the end of the wire, slip the rod into the loop, and pull upward to rotate the anchor 90° in the hole, locking it into place. Use the loop to attach your guy.

Duckbill anchors come in various sizes and holding strengths. Their one drawback is that, unlike screw anchors, they cannot be reused.

Foots

A foot is a wedge block used to increase the stability of any anchor post set in a hole and back-filled with relatively stable soil. It is especially useful for strengthening a dip post or the post of a diagonal stay assembly. It can also be used in high-tension stranded wire fences to keep the line wire from twisting end posts in their holes.

The foot consists of a pressure-treated 2x2 (5x5 cm), 12 inches (30 cm) long, with one end cut at a slant. A length of 8-gauge galvanized soft wire is wrapped around the center and stapled to the foot. The footed post should be 6 inches (15 cm) in diameter and set in a 12-inch-wide (30 cm) hole, at least 3 feet (90 cm) deep.

Wedge the foot into the hole beside the post, pointing along the fence line. Wrap the wire one-half turn around the post, tie a loop at the end, and staple the wire against the post approximately 8 inches (20 cm) above ground level.

If you are footing a post to keep it from being twisted by taut line wire, wrap the foot's wire in the direction the line wire will pull. If your fence will have more than five wires, you will need two foots (not "feet"), one on each side of the post.

LINE POSTS

Just as proper bracing is an important aspect of anchor posts, proper spacing is an important aspect of line posts. Different kinds of fence have different spacing requirements. Since every fenced area is

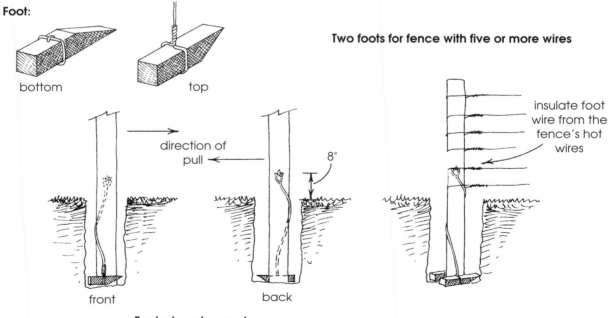

Foot:

bottom   top

Two foots for fence with five or more wires

direction of pull ←

8"

insulate foot wire from the fence's hot wires

front   back

Footed anchor post

unique in size, shape, and terrain, ideal spacing can often be difficult to translate into reality. If you start spacing posts at one end, chances are you'll wind up with an odd span at the other end.

If that odd span falls in some back corner where no one will notice, great. But an odd span at the front of a fence or beside a gate makes it look like something went wrong. A careful fence builder will juggle the spacing to avoid this sloppy look.

When post spacing doesn't come out exactly even, you can add an extra post and space all the posts closer together. Suppose, for example, you want to space line posts 16 feet (5 m) apart along a 120-foot (36 m) run of fence. Dividing the second number by the first gives you 7½, an impossible number of posts. If you round that up to 8 posts, then divide into 120 feet (36 m), you'll come out with an even 15 feet (450 cm) between posts.

Rail fence builders don't like this method because it requires shortening each rail. Instead, they prefer to divide the odd span in half and add a short span at each end of the run, next to the anchor posts. Another alternative is to put the odd span smack in the middle of the run and add ornamental boards to form crosses or diamonds, making the odd span look as if it was deliberately designed into the fence.

The designated line post spacings for each kind of fence represent ideal conditions: level terrain, temperate climate, no strong wind. Closer spacing may be necessary because of rough terrain, strong prevailing wind, or cold weather. In Western areas, wind causes blowing weeds to pile up against a fence. In the North, ice storms, snow slides, and blowing drifts add to a fence's load.

If such conditions are likely in your area, for anything other than a suspension fence, space wood posts no more than 20 feet (610 cm) apart, steel posts no more than 14 feet (425 cm). Other situations that require closer post spacing are curves, steep grades, and dips or rises.

Post Alignment

For a corral or rail fence in which the fence material goes through the center of each post, line up posts on their centers. For a wire or rail fence with fence material affixed to the side of each post, align posts on that side. Side alignment is especially important if you're setting wood posts, which can vary considerably in diameter. Use a carpenter's tape measure or a length of carpenter's string, knotted at appropriate intervals, to measure the distance between posts. Always measure from the center of one post to the center of the next.

For short runs of fence, tie a string between adjacent anchor posts and use it as a guideline. The string should be about 6 inches (15 cm) off the ground, or as high as necessary to avoid snagging rocks, weeds, or clumps of dirt. Mark the position of each post by making an "X" with agricultural lime, spray paint, or corn meal. If you are working in grass where such marks won't show, substitute pegs flagged with surveyor's tape. If you are stringing a wire strand fence, stretch the bottom wire first and use it as your guide. Set your line posts a fraction of an inch away from the wire so your fence

LINE POST SPACING

Fence	Post Spacing	
Barbed wire	12 to 14 ft	360 to 420 cm
High-tension	30 ft	915 cm
Electric, conventional	12 to 14 ft	360 to 420 cm
Electric, high-tension	40 to 100 ft	1,200 to 3,050 cm
Woven wire, conventional	16 ft	480 cm
Woven wire, high-tension	24 ft	480 cm
Reel	10 ft	305 cm
Corral	5 to 8 ft	150 to 240 cm
Rail	8 to 10 ft	240 to 300 cm

doesn't start wandering off into left field.

For long runs of fence, where a guide string or wire may not be practical, you will need a partner to help you accurately align your posts. On level terrain, have your partner establish the location of every tenth post while you stand at one anchor post and sight their alignment with the distant anchor post. Once you have these posts properly aligned, run a string from one to the next to establish the positions of intervening posts.

On hilly land where you can't easily see one anchor post from the next, set two sighting poles no less than 10 feet (3 m) apart at the top of the intervening hill or at the bottom of the intervening valley, so both poles can be seen from both anchor posts. Working with your partner, line up the two poles from one anchor post, then from the other. Keep checking back and forth until the sighting poles line up from both ends.

If you are working alone, run a string from one anchor post to the top of the pole farthest from it. Run a second string from the other anchor post to the top of the pole farthest from it. Adjust the two poles until each just touches the other's string. Set a rise or dip post at the highest or lowest point, and use it to align your line posts with the anchor posts on either side.

Curves

For a wire fence, straight runs are ideal because they place the least amount of pressure on line

Aligning line posts

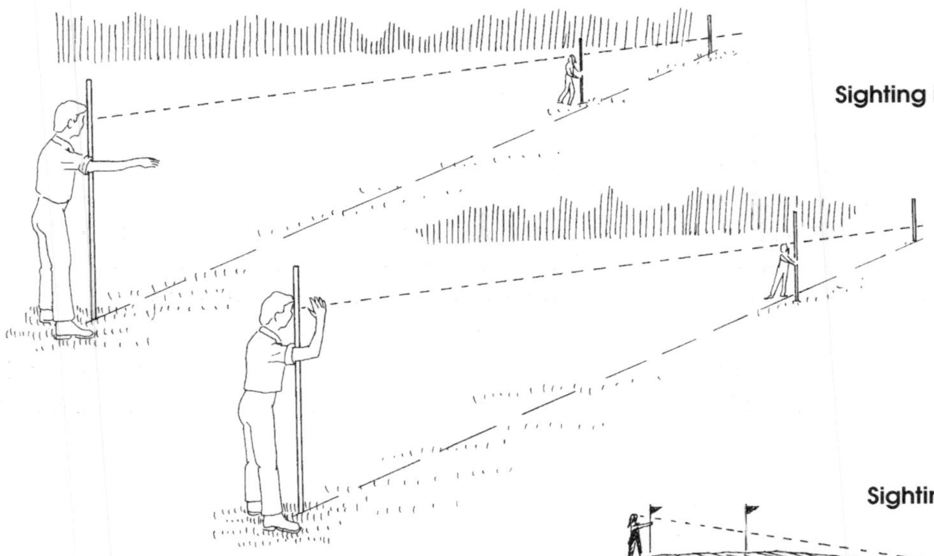

Sighting in on level terrain

Sighting in on hilly terrain

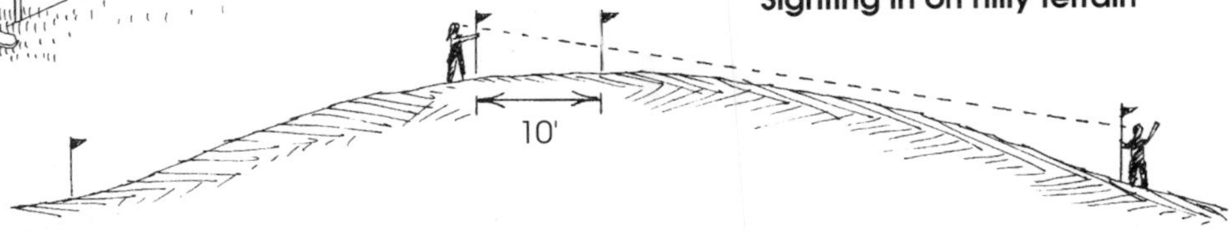

10'

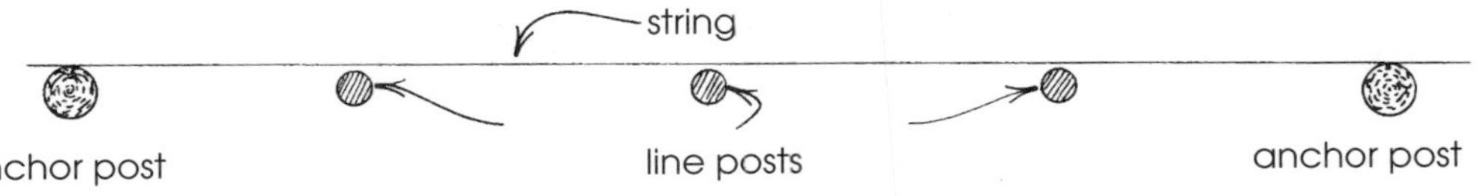

string

anchor post   line posts   anchor post

Don't let posts touch the alignment string or fence may wander to one side.

posts, but sometimes you can't avoid fencing on a curve. The fence might follow a contour line, for example, or run along a winding roadway. You might deliberately round off your corners, for example to make them easier to mow. If you raise sheep, you know they tend to follow the fence line and put extra pressure on corners, something you can prevent simply by curving the fence to eliminate corners. Rounding off tight corners of an electric fence is an especially good idea, since animals won't go into such corners to graze lest they get zapped, and you won't want to mow there for the same reason.

Curve posts have more pressure on them than line posts in a straight row, but you can avoid adding brace assemblies if you arrange the posts so the fence never changes direction more than 20° at one post. You don't even have to calculate each angle of change. Use the following method of linear measurement involving stakes, string, and a measuring tape.

Space the stakes 16 feet apart along the line of the curve and weave the string in and out between them. Measure the distance between the string and each intervening stake, and use the accompanying chart to determine how far apart the posts need to go along that part of the curve. Unless the curve is a smooth, broad sweep, you will likely need to space posts closer together at some parts than at others, so measure every stake.

After adjusting the stakes according to the chart, run the string around the outside of all of them to see if they fall in a smooth arc. Smooth out the curve by readjusting any stakes out of alignment until the string touches every one.

If the curve is so sharp that the chart calls for posts closer together than 8 feet (2,440 mm), it is much easier and no more expensive to install H-brace assemblies around the curve rather than adding more posts. In any case, never space posts closer together than 4 feet (1,220 mm), since doing so disturbs the soil enough to reduce its holding power.

Curve posts should be stouter than line posts, set deeper into the ground, and tilted outward to compensate for the inward pull of wire. How much tilt you need depends on the degree of curve: 2 inches (5 cm) is enough for a shallow curve; 5 inches (13 cm) is more appropriate for a deep curve.

If a curved fence line undulates, place a pull-post assembly at the start and end of each curve, as illustrated. When tightening wire around a curve, start from the end with the least curvature so the part with the most curvature experiences the least amount of tension.

CORRALS

Laying out a circular corral is a bit different from laying out a curved fence line. Here you will need a bundle of pegs, one round metal stake, and one 8-foot long (240 cm) 2x4 (5x10 cm) less the thickness of one peg and with a hole drilled near each end.

You will also need a length of stout twine, slightly

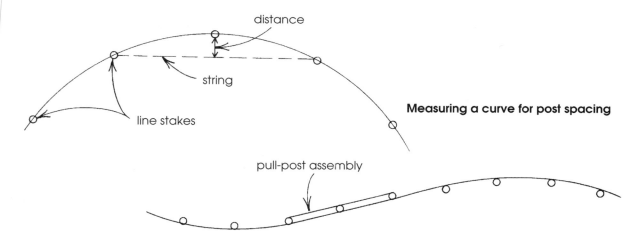

Measuring a curve for post spacing

For an undulating curve, place a pull-post assembly at the beginning and end of each curve.

longer than twice the radius of your corral, with a loop in the middle and the two ends tied to the holes in the 2x4 (5x10 cm). The distance between the loop and the middle of the 2x4 (5x10 cm) will be the radius of your corral. The smallest corral you can lay out by this method has a radius of 35 feet (1,070 cm).

Securely position and plumb the metal stake at the center of your proposed corral and slip the loop around it. Lay the 2x4 (5x10 cm) on the ground with the string fully stretched, starting where you want to put your gate. Place a peg in the ground at each end, thus marking the positions of the first two posts. Work your way around the circle, setting out

SPACING POSTS ON A CURVE

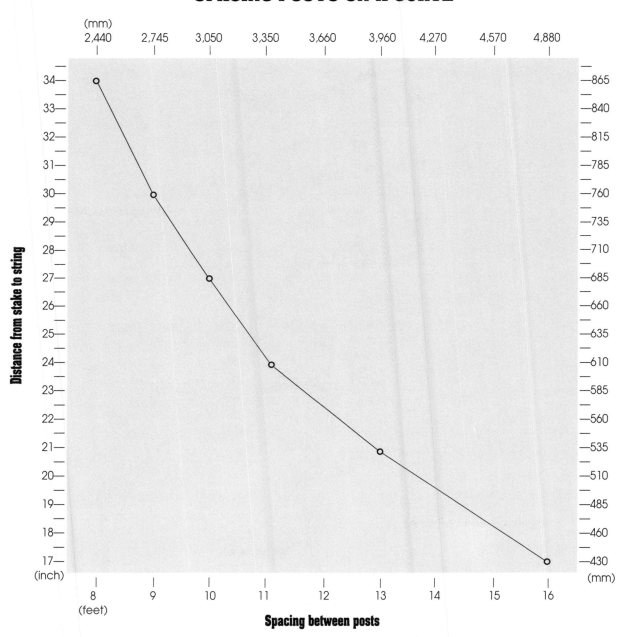

Distance from stake to string

Spacing between posts

If distance is less than 8 feet (2,440 mm), place horizontal braces between posts.

post marker pegs as you go and keeping the twine tight and the 2x4 (5x10 cm) snug against the previous peg.

When you get back to the first marker, one of three things will happen. Either post spacing will come out exactly even and the last span will be just the right size for an 8-foot (240 cm) gate (possible), or the last span will be too wide or too narrow for your gate (likely). If the opening is too wide for your gate, add a short span of fence on either side to narrow it. If the opening is too narrow, widen it by adjusting the posts to shorten the spans on each side by an equal amount.

Use this same method to round off a square corner in a large field. Pace off a distance of 35 feet (1,070 cm) from the corner down each side of the fence, then pace off the same distance at right angles to find the spot diagonally opposite the corner post on a 35-foot (1,070 cm) square. Position the stout stake at that spot. Loop the string over it, stretch it to a 35-foot (1,070 cm) radius, and measure off your fence posts as before.

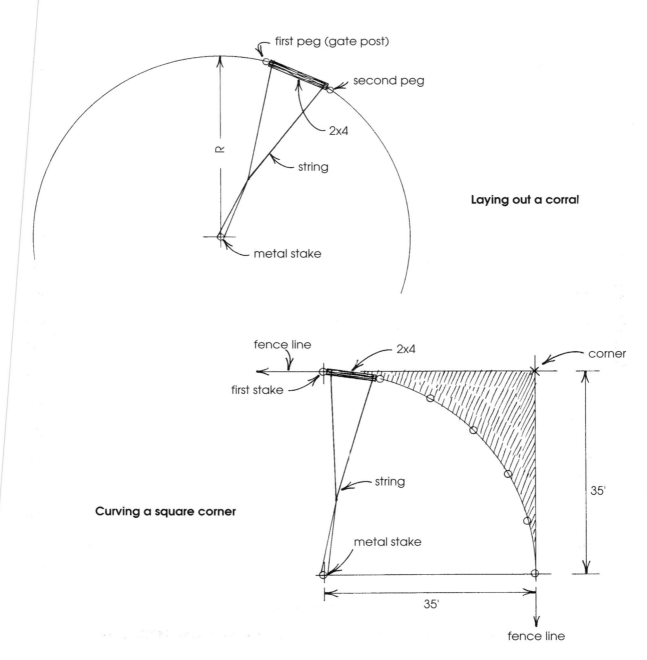

Laying out a corral

Curving a square corner

CHAPTER 4 ▾▴▾▴▾▴▾▴▾▴▾

SETTING POSTS

▾▴▾▴▾▴▾▴▾▴▾▴▾▴▾▴▾▴▾▴▾▴▾

When you see a fence leaning in all directions, it means the builder didn't set the fence posts deep enough. How deep your posts need to go depends on the kind of fence you're building, the nature of your soil and terrain, the climate and depth of frost penetration in your area, and whether you're setting line posts or anchor posts. The method you use to get the posts into the ground depends on the kind of posts you're setting and on your soil type.

Before putting in your first post, check with local utility companies to make sure you won't cut into underground pipes or cables. You could be held liable for damage unless you've been given the go-ahead by a utility representative. To find out whom to contact, consult the "one-call" service listed in the appendix.

POST DEPTH

The deeper you set your fence posts, the better they will hold. Because setting posts is hard work, most amateur fence builders give up long before they get deep enough. Pressures caused by frost heave, fencing material weight or tension, and back-rubbing animals eventually cause the posts to wobble and the whole fence to sag.

Agricultural engineers have devised a set of formulas to determine how deep posts should go based on their height and diameter, the nature of the fence, and the soil type. Short of hiring an engineer to analyze your particular situation, you

can follow certain minimum standards. If rocky soil or hardpan prevents you from setting posts to their minimum depth, put them in concrete.

Anchor posts and brace posts incur much more stress than line posts, so they need to be set deeper. Whether or not a gate post should go deeper than other anchor posts depends on the weight it must support. The heavier the gate, the deeper the post must go. Any post in sandy or swampy soil must be set deeper than a post in firm soil. A post set in well-drained soil with strong holding power can bear a fence load two or three times greater than the same post set in low-strength, poorly drained soil. You can double resistance to pull-out by increasing a post's depth 6 inches (15 cm) in firm soil or up to 18 inches (45 cm) in less than ideal soil.

Under normal conditions, set wood anchor and brace posts for a low-tension fence 3 to 3½ feet (90 to 110 cm) deep, fiberglass O-posts 3 feet (90 cm) deep, and steel posts 2½ feet (75 cm) deep. In northern climates, all posts must go below the frost line to avoid frost heave.

For a high-tension fence, the minimum depth for anchor posts is 3 feet (90 cm), ranging up to an extreme of 8 feet (240 cm) where the soil consists entirely of mud or loose sand. Lean high-tension anchor and brace posts 1 to 2 inches (25 to 50 mm) away from the direction of pull so they'll straighten up when the fence is tensioned.

Set curve, rise, or dip posts as deep as you would any anchor post. For a wire fence, lean the tops of curve posts 2 inches (5 cm) toward the outside of the curve — they will pull into line when you tension the wire.

Line posts aren't subject to the same strain as anchor posts, so they needn't be set as deep. One exception is the posts for a corral or barn lot, which should be set 3 feet (90 cm) deep to accommodate heavy animal pressure. Another exception is rail fence posts, which also need to be 3 feet (90 cm) deep to hold up under the weight of rails. Otherwise, set wood posts at least 2½ feet (75 cm) deep, steel or fiberglass posts 1½ to 2 feet (45 to 60 cm) deep.

Make sure all your posts are long enough to accommodate these minimum depths plus the height of your fence plus a 4- to 6-inch (10 to 15 cm) leeway at the top. To ensure all posts are set to their

MINIMUM POST DEPTHS

Anchor Posts

wood, LT	3–3½ ft	90–110 cm
wood, HT	4 ft	120 cm
fiberglass	3 ft	90 cm
steel	2½ ft	75 cm

Line Posts

rail/corral	3 ft	90 cm
wood	2½ ft	75 cm
fiberglass/steel	1½–2 ft	45-60 cm

proper depth, mark appropriate ground level on each with a carpenter's crayon.

If your fence goes over hill and dale, set posts *perpendicular* to the ground, not plumb. Plumb posts may look better, but they increase the difficulty of keeping the fencing material at an even height and close to the ground. A gap at the bottom becomes a detriment if one of your goals is to discourage diggers and crawlers. Two exceptions are trellis posts, which are always set plumb, and the posts for a stepped rail fence, described in chapter 12.

TOOLS AND METHODS

You can set posts in one of two ways: pound them into the ground or plant them in holes. Steel, fiberglass, and plastic posts are usually driven. Stone, concrete, vinyl, and vinyl-clad posts must be hand-set in dug holes, since driving would mar their appearance. Any post with a deadman requires a dug hole. So does any post requiring a concrete footer. Wood posts can be either hand-set or driven — whether or not you have a choice depends on how hard or rocky your soil is.

Post holes can be dug manually or mechanically. Manual digging is difficult in anything but damp or fairly soft soil. In hard-packed or rocky soil, you will need a mechanical auger. The holes can be back-filled either with stones and soil or with concrete. Using concrete is the most expensive method of setting a post, but it also offers the firmest hold. Backfilling with stones and dirt is cheaper, but results in the weakest hold. As a general rule, hand-set line posts may be backfilled with soil but hand-set anchor posts in loose soil should be set in concrete.

Driving posts into the ground is the fastest method and the second strongest. Posts drive more

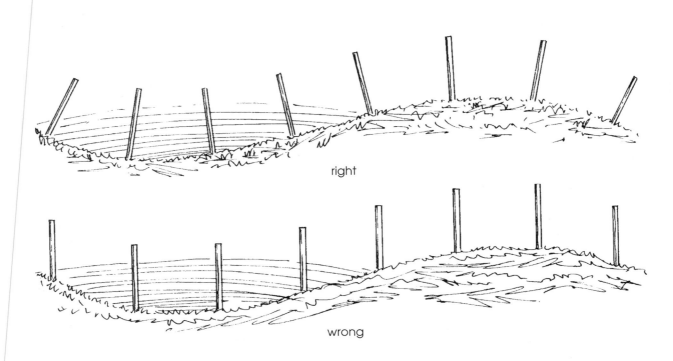

right

wrong

Setting fence posts on a hill

easily in wet weather, but also pull out more easily. If you drive posts during the rainy season and you are building a wire fence, wait until the soil dries out before stretching the wire.

Sometimes a combination is your best option — drilling a hole that's narrower than the post and then driving the post in, or drilling a pilot hole and driving the post the rest of the way. Starting pilot holes is an especially good idea for reducing damage to the tops of fiberglass posts driven into hard or crusty soil.

Instead of buying expensive post-setting equipment, you can lease it from a rental agency. In some areas farmers or contractors will do the job for a small fee per post. Since it is not wise to make more holes than you can fill the same day, if you have post holes custom dug, assemble a crew to help you get your posts in.

Manual Diggers

Unless you're digging only a few holes in loamy soil, using a shovel can be an excruciating form of torture. Furthermore, with a shovel you tend to dig a hole that's wider at the top than at the bottom. A post hole should be straight up and down or, where frost heave is a problem, wider at the bottom than at the top. Besides, it is impossible to dig deep enough with a shovel unless you make the hole much wider than you need.

If you are determined to dig holes by hand, use a proper post hole digger. Along with the digger, you will need a long, heavy pry bar to loosen the soil

Using a digging bar   **Using a clamshell digger**

and to pry out or break up rocks that get in your way. Two kinds of pry bar are made specifically for fencing. One, called a "digging bar," is pencil-pointed at one end and has a crowbar-like wedge at the other end. Use the pointed end for digging and the wedge end to pry out rocks. The other kind, called a "digger/tamper," has a flattened blade at one end and a squared-off tamper at the other. Use the blade for digging and prying, the tamper for packing soil when you backfill.

If you are setting round, thin posts made from rebar or fiberglass, and your soil is too hard to pound them in, you can often drill holes with the pry bar. Raise the bar and let the pointed end fall into the soil, then rotate it to create a conical hole. Keep this up until the hole is deep enough, set the post at the center, then backfill and tamp the soil smooth. If you are setting larger posts requiring deeper, wider holes, and your soil is hard or rocky, use the pry bar to loosen the soil, then scoop it out with the post hole digger.

A standard post hole digger, sometimes called a "clamshell" digger, has two parallel handles and two hinged blades, shaped like clamshells, that dig a perfectly cylindrical hole. In theory, you're supposed to lift the digger and let it drop under its own weight. In reality, unless your soil is quite loose, you can't avoid slamming the tool into the ground (and rapidly wearing yourself out). Spread the

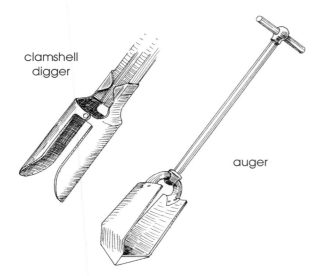

clamshell digger

auger

handles to bring the blades together and trap a load of dirt, lift the digger out of the hole, and set the dirt far enough away that it won't slide back into the hole.

The standard spread of a digger designed for home use is 5½ inches (14 cm). Larger, more expensive models are used by fencing professionals. Avoid el-cheapo diggers, sold to unwary folks who prowl discount stores. Because these tools don't hold up well, they are considered "disposable" by those who manufacture them. They can be identified by their short, lightweight wood handles and unsharpened blades.

For serious fence building, you will want a heavy-duty digger with sharpened, carbon steel blades and handles at least 58 inches (145 cm) long, made of number one grade hardwood, welded steel pipe, or high-strength fiberglass. If your digger is destined for hard use, consider one with replaceable blades. The blades require periodic sharpening so they can cut through small roots. If you encounter large roots, cut them with pruning sheers, a sturdy knife, or a digger/tamper bar.

How deep you can go with a manual digger depends on the length of the digger handles. Another limitation is the increasing difficulty of spreading the handles the deeper you go, unless you widen the hole. Widening can be accomplished by chipping at the sides of the hole with the digger/tamper bar, then scraping the sides smooth with the digger blades. Some diggers have marks on their handles for quick depth reference. If yours doesn't, you can easily mark either your digger handle or your pry bar.

A modified clamshell digger, designed for use in gravel or rocky soil, has one normal handle attached to a fixed, straight blade. The second handle has a lever-operated hinge and is attached to a curved blade. Instead of pushing the handles together to separate the blades and pulling them apart to bring the blades together, you pull the lever up to open the blades and push it down to trap dirt. With this device you can dig until the hinged lever hits ground level, or about 4 feet (120 cm).

Another variation is the manual auger with a right-angle handle or "twist bar" at the top. Instead of dropping the blades into the soil, you push them in, then lean against the bar and rotate. Some

models are adjustable so you can dig holes varying in diameter from 5 to 7 inches (13 to 18 cm).

Power Augers

A power auger is faster than a manual digger and is more suitable for drilling holes on slopes and in hard-packed or rocky soil. Power augers come in numerous forms including self-contained units and attachments for chain saws or tractors.

A power auger looks like a giant corkscrew and operates on much the same principle. Auger bits come in a variety of sizes ranging upward from 2 inches (5 cm) in diameter. Auger shapes also vary to accommodate different soil conditions including solid rock, frozen ground, hardpan, and sand. If you are digging into sand or other loose soil that tends to cave in, there is even a hollow bit through which you can insert a post before retrieving the auger.

Self-contained units, usually gasoline driven, are either portable or trailer-mounted. Some portable units are small enough for one person to operate, others require two. A one-person auger is smaller and lighter, has less power, and will drill holes up to 8 inches (20 cm) in diameter. A two-person auger drills faster and will make holes as wide as 16 inches (40 cm). Avoid the temptation, and the risk, of trying to operate a two-person auger by yourself.

In rocky or root-filled soil, you need a large build and a strong back to run a portable auger. At best, a portable auger in any soil is rough on your arms, legs, and lower back. To reduce back strain, bend

A one-person portable power auger with detached power source

your knees as the auger digs down. Should the rotation suddenly be stopped by a rock, root, or other hard obstacle, be ready for one heck of a kickback.

A portable model with a detached power source is lighter and therefore easier to handle than a unit with the power source directly above the bit. By winching the auger head from a tripod, you can further reduce the need for muscle power.

Larger, trailer-mounted augers are more powerful than hand-held ones and can be towed behind any vehicle. Here, the auger head is supported by a sturdy stand. Once you set up the equipment, operation largely entails pushing buttons and pulling levers.

If you already own a suitable power source, you might opt for an auger attachment. The kind that fits your chain saw might be tempting, but it will

A two-person power auger with power source on top

drill holes of only limited diameter. Besides, if you drill a lot of holes, you will soon wear out your saw's chassis and motor.

If you have a tractor, buy or rent an auger attachment that operates mechanically or hydraulically. A mechanical drill attaches to the three-point hitch and derives power and digging speed from the power take off (PTO). Using one of these, you can

drill holes up to 24 inches (60 cm) in diameter and 6 feet (180 cm) deep — which covers just about any fence post hole situation.

A hydraulically driven auger is operated by a tractor's auxiliary hydraulic system (or any other hydraulic source). It will drill holes to 48 inches (120 cm) in diameter and 10 feet (300 cm) deep or more — slight overkill for most home fences.

Before operating any power auger, read the instruction manual and watch any safety video provided. Use gloves to improve your grip and wear sturdy clothing with snug-fitting sleeves and pant legs. Leg padding provides additional safety. Stop the auger when you move it from one hole to the next. If you add an extension for a deeper hole, attach it only *after* the hole has been partially drilled. Finally, when digging in rocky soil, keep a few replacement teeth, as well as mounting bolts and nuts, on hand. It will save you a trip to the store when (not if) an auger tooth breaks or a bolt wears through.

Post Holes

Dig post holes straight down and wide enough for posts to be properly placed and plumbed. As a general rule, make holes at least twice the diameter of your posts. An irregular hole or one with too little space around the post makes firm backfilling difficult.

When setting wood posts in poorly drained soil, dig 6 inches (15 cm) deeper than the post will be set so you can fill the bottom with coarse gravel or small stones to improve drainage and slow down wood decay. You won't need extra gravel if you are setting a post in soft or loose soil that naturally drains well, but you can strengthen the post by placing a large, flat rock in the bottom of the hole.

Set a wood post with the large end downward. Check two faces of all posts with your carpenter's level before you start backfilling. Since it's easy to bump a post while you work, keep checking that the post is plumb until it is secure.

According to folklore, if you dig a hole on the shrinking side of the moon, you won't have enough dirt to refill it. If you dig on the growing side of the moon, you'll have more dirt than you know what to do with. No one has ever verified this scientifically, but if you dig enough post holes you'll find that sometimes you won't have enough dirt to backfill while other times you'll need a wheelbarrow to cart away the excess.

Actually, backfilling with dirt is not a good idea, but you can get away with it if you have the kind of soil that packs down hard. At the least, add a shovelful of gravel or a few small rocks around the bottom of the post to hold it in place. Better yet, fill the hole with gravel to within 6 inches (15 cm) of grade and you'll greatly increase the post's holding power.

Fill the rest of the hole with soil, adding only a few inches at a time and tamping well before adding more. Pack the soil as close to its original density as possible. How firmly the post will hold depends entirely on how firmly the soil is tamped. Use the end of a 2x4 (5x10 cm) or a 2-inch-diameter (5 cm) wood pole or fiberglass rod for tamping; a heavy steel tamper works even better.

To settle soft soil or sand, pour a bucket of water around the post after it has been set. Finish by grading around the post so the soil gently slopes away on all sides, thus reducing the chance that standing water will seep in around the post.

Don't be tempted to dig all your post holes in one day and set all your posts the next. Dig only as many holes as you can fill in the same day. Holes left overnight may collect rain water, making backfilling more difficult. Besides, post holes have a mysterious way of caving in. Redigging is not only more work, but invariably results in a hole of irregular shape. Irregular holes can be a real problem if you're setting posts in concrete.

Setting Posts in Concrete

In loose, moist, or sandy soil, or in hard soil that doesn't allow you to dig deep enough, posts hold better if you set them in a footing of concrete. Corral posts and rail fence posts in anchor positions, as well as the line posts next to them, should always be set in concrete. All posts for a chain link fence should also go into concrete. So should every gate post.

Setting posts in concrete *isn't* necessary for low-tension wire fences or for wood posts in clay loam. It's downright undesirable for nontreated wood posts; they tend to shrink and leave a moisture-collecting crack between the post and the footing.

For a concrete footer, make the post hole at least twice the diameter of a stout post, up to four times the diameter of a thinner post such as those used for chain link fences. Plant posts at least 2 feet (60 cm) deep for a 4-foot-high (120 cm) fence. Add 3 inches (8 cm) for each additional foot (30 cm) of fence height. As you dig, move loose soil away from the sides of your holes so none can accidentally slide in and weaken your concrete. Moving away loose soil also lets you more easily see grade level while you're pouring.

To minimize frost heave, keep hole sides as smooth as possible — irregularities give frost a foothold to push against. If frost is a serious problem, enlarge post holes at the bottoms by either flaring them into bell shapes or rounding them into balls. The larger the footing, the more mass there will be for the frost to contend with.

Post holes

straight   bell-shaped   ball-shaped

To improve the bond between a wood post and its concrete footer, hammer a few stout galvanized nails perpendicularly into the lower part of the post, just deep enough so they'll hold firmly. For a vinyl or concrete post, insert an 18-inch (45 cm) length of rebar into a hole at the base. For steel pipe posts, have a welder fix short lengths of rebar to the bottoms — the cost will be minimal.

Place gravel in the bottom of each hole to improve drainage. Put a flat rock on top of the gravel so wet concrete won't filter down through the gravel. Center the post and fill around it with tamped-down concrete. Avoid getting concrete beneath the post. Otherwise, when it hardens it will form a bowl around the post that will trap water and hasten decay or rust. Furthermore, should the water freeze, the concrete would crack.

As you work, keep checking two faces of the post with a level. If you're working with a partner, one of you can hold the post in place while the other pours the concrete. If you're working alone, take time to brace the post so it can't shift — it will be too late to adjust the post after the concrete starts to harden.

Use a 1-2-4 mixture (one part cement, two parts sand, four parts gravel). If you prefer to use a premix, add four shovelfuls of gravel to each 80-pound (36 kg) sack. One 80-pound (36 kg) bag will fill a ⅔-cubic-foot (1.8 cu cm) hole. Premix also comes in 40- and 60-pound (18 and 27 kg) bags in the U.S.; in Canada, it is packaged in 40-kilogram (88 lb) bags.

Keep the mix a bit on the dry side. Soupy concrete is too weak to hold a post plumb while it hardens and is more likely to crack after it does harden. With a soupy mixture you will also have trouble crowning your footings, which you should do by heaping concrete around the post, then sloping it outward on all sides.

Crowning with concrete improves drainage and keeps weeds from growing around the post. If you prefer not to see the finished footing, or if you're following the common practice of adding one sack of premix per hole and it doesn't reach the top, the day after pouring your footers, finish filling the holes with soil. Heap soil around the posts, then

CONCRETE REQUIREMENT FOR POST HOLES

Hole Diameter		Hole Depth				
		24 in.	30 in.	36 in.	42 in.	48 in.
8 in.	C	.70	.87	1.05	1.22	1.4
	P	38.5	31	25.7	22	19
10 in.	C	1.1	1.36	1.64	1.91	2.1
	P	24.5	19.7	16.5	14	12
12 in.	C	1.57	1.96	2.36	2.75	3.14
	P	17.5	13.7	11	9.7	8.5
14 in.	C	2.13	2.67	3.21	3.74	4.28
	P	12.5	10	8	7	6

C = cubic feet of concrete per hole
P = number of posts per cubic yard of concrete

METRIC CONCRETE REQUIREMENT FOR POST HOLES

Hole Diameter		Hole Depth				
		610 mm	762 mm	914 mm	1,067 mm	1,219 mm
203 mm	C	.020	.025	.030	.035	.040
	P	50	40	33.3	28.6	25
254 mm	C	.031	.038	.046	.054	.059
	P	32.3	26.3	21.7	18.5	16.9
305 mm	C	.044	.055	.067	.078	.089
	P	22.7	18.2	14.9	17.8	11.2
356 mm	C	.060	.076	.091	.106	.121
	P	16.7	13.2	11	9.4	8.3

C = cubic meters of concrete per hole
P = number of posts per cubic meter of concrete

gently slope it away and plant grass to choke out weeds.

Mixing concrete can be a hassle if you're away from a ready source of water. An easier method is to use fast-setting concrete. Pour dry fast-set into the post hole and add a bucket of water. The concrete will set in a matter of minutes. This way you won't have to haul extra water to rinse out your trough and tools as you work. Even easier is dry-set concrete, which requires no water at all. Use it to form a collar around the post at its base, cover the dry mix with tamped-down dirt, and add a second collar 1 foot (30 cm) below the surface.

Let concrete footers cure at least two days before continuing construction. Let a gate post footer cure seven days before you hang the gate.

Post Drivers

By far the best way to set a post is by driving it into the ground. Since driving compresses the soil, a driven post has considerably more holding power than a hand-set post. For the strongest fence, therefore, drive your fence posts whenever possible, especially those in anchor or brace positions.

Steel and fiberglass posts are almost always driven, usually with a manual driver. Steel posts must be perfectly straight before you start or driving will bend them even further. Wood posts are more difficult to drive, especially if they're stout or the soil is hard, making a power driver more suitable than a manual driver.

Manual Drivers

If you're using plastic or fiberglass posts for a temporary fence, you should be able to push them into loose soil with your bare hands. If the soil is hard-packed or rocky, you will need a hammer or mallet. A 2-inch (5 cm) rubber mallet does the least amount of damage to post tops. If you use a regular hammer, prevent posts from splaying by topping them with a block of wood or a protective cap. Caps are available from suppliers of fiberglass posts.

It is possible — though neither particularly safe nor sane — to drive steel posts or pencil-pointed wooden ones with a sledge hammer or fence maul. A fence maul, with its short, heavy head and rounded faces, reduces the chance that you'll split a wood

post if your aim is off. Whether driving wood or steel posts, wear safety goggles in case of flying splinters. And don't put anyone in the dangerous position of holding a post steady while you hit it. Instead, use temporary bracing.

Before you start, make sure the hammer head isn't loose and the handle is sound — a flying head can be lethal. Hit the post at approximately waist height, which means you may need something to stand on. The bucket of a front-end loader makes a solid platform that can be raised and lowered as necessary, but the tailgate of a pickup truck also works. Don't be tempted to stand on a ladder or anything equally tippy. Face the post squarely and spread your feet, both for good balance and so you won't whack your shin if you miss the post.

Drive fiberglass posts with a rubber mallet (left), drive metal posts with a tubular post driver (right).

Far better than a maul for manual driving of steel, stout fiberglass, and wood line posts is a tubular post driver. The simplest version consists of a steel tube, 2 feet (60 cm) long and 5 inches (13 cm) in diameter, capped at one end. As a cushion, an oak block may be inserted into the capped end. Commercial models have handles welded to opposite sides, making the driver easier to lift and slam down. Some are spring-loaded to save you a bit of muscle power. When you hit a post, the spring causes the driver to lift, hit again, then partially lift for the next stroke.

Start a leader hole with your pry bar, drop the post in, slip the driver over the top of the post, and pound away. If you're setting more than a few posts, you will find manual driving to be slow, painful, exhausting work, especially if your soil is not loose or loamy. Manual driving is difficult in hard soils and impossible for wood posts much over 4 inches (10 cm) in diameter. Power provides a faster, less taxing method for setting posts.

Power Drivers

Like power diggers, power drivers come either as self-contained portable units or as tractor attachments. A tractor-mounted driver connects to the three-point hitch. It works by hydraulic lift and free fall, making up to forty-five strokes per minute. It will drive steel posts or wood posts up to 8 inches (20 cm) in diameter. Most portable units drive only standard steel posts or pipes.

Like a power digger, a tractor-mounted driver lets you set posts straight, even on a slope. You can adjust for only 5° of angle; on steeper slopes, you're out of luck. Like most power diggers, a power driver doesn't work well in extremely rocky soil.

Even when you use a driver, you have to start pilot holes to get your posts positioned properly. To set extra-tall posts, you will need to dig or drill holes deep enough to get the driver started. When using a portable driver, mark the soil line on each post so you will know when to stop. A tractor attachment has automatic depth control.

Follow the same safety precautions for power driving as for power digging, and wear ear plugs to protect your hearing. If you're driving wood posts larger than 4 inches (10 cm) in diameter, sharpen the ends to a dull point. Drive wood posts small end down — exactly opposite of the way you'd set them in dug holes.

SETTING POSTS IN WATER

In marshy or boggy land where digging is difficult, posts must be driven. Even so, they will never be as stable as posts in firm, dry soil. If setting posts in wet soil is unavoidable, improve stability by using extra-long posts and driving them deeper. If the soil consists of sand or loose silt, add concrete footers.

You can pour concrete under water if the water temperature is at least 45°F (7°C). Mix the concrete with only enough water to moisten all particles. To shape it, you will need a form such as a large bucket or drum with the bottom removed, a short length of stout culvert, or a sturdily nailed-together wood form, preferably wider toward the bottom. If the top of the form reaches the water's surface, you won't need a funnel to channel wet concrete into the form.

When setting steel posts, the so-called "sack" method eliminates the need for forms. Simply fill jute sandbags two-thirds full with dry concrete mix. Drive a post into place, ram a sack over the top, push it to the base of the post, and tamp it down. For best stability, allow two sacks per post. Concrete from one sack will seep into the other, bonding the two together.

When a line post falls in a gully where run-off periodically flows, why take a chance the post will

APPROXIMATE* TIME NEEDED TO SET A POST

Kind of Post	Method	Setting Time
Wood/vinyl/stone/concrete	Hand dug and set	25 min.
Wood	Power dug, handset	12 min.
Wood	Power driven	4 min.
Steel/fiberglass	Hand driven	8 min.
Steel	Power driven	1 min.

*Actual time varies with post depth and soil conditions.

get washed out? Instead, replace it with a pair of posts, one on each side of the gully. If the gully is more than 16 feet (5 m) wide, use stout posts braced with anchor assemblies and stretch a separate run of fence between them. If a flood washes away your fencing material, you will have to replace only the short section between the assemblies. A full discussion on running a fence across water gaps appears in chapter 13.

POST PULLERS

At one time or another nearly every fencer has to pull up posts, either to relocate a fence that is in the wrong place or to tear down and rebuild an old one. Begin by removing any attached fence wire. In all probability it won't be worth saving, but taking it up makes post pulling easier and safer. Take care, though, since loose fence wire has a way of whipping around and snagging you when you least expect it. For safety's sake, try to work from the side of the post away from the wire. Carry a bucket to hold short pieces of wire and old clips or staples.

The trick to pulling a post is to yank straight up without bending the post (if it's steel) or breaking it off at ground level (if wood). If the post isn't set very deep, loosen it by wiggling it back and forth. Then, with your knees bent, get a firm grasp and pull upward. If the post is set in concrete or hard soil, or goes quite deep, you will have a hard time getting enough leverage to loosen and pull it by hand without incurring back strain.

In that case, wiggle it loose by wrapping a cable around the top. Attach the other end to a come-along or vehicle bumper, and apply gentle pressure until the post moves out of plumb. Repeat from different directions and the post will soon be loose enough to pull. If it is set in concrete, first remove some of the dirt around the base.

Another way to pull posts without pulling muscles is to use a high-lift mechanical jack. Bolt a slip hook to the jack's lift plate, hook into a chain wrapped around the post, and crank away. If you're pulling T-posts, buy a plate that slips over the post and hooks onto the jack, or buy a jack designed especially for pulling T-posts.

Another alternative is to rent a self-contained, one-person hydraulic post puller. Wrap a chain around the post, attach the puller, and up it goes.

Tractor owners have all sorts of post-pulling alternatives. Wrap a tow cable around the post near ground level and use the hydraulic lift or three-point hitch to pull the post up. If the post is set in concrete and the tractor has a bucket loader, loosen the post first by pushing gently against the top with the edge of the bucket. If you want to get fancy, use a three-point post-pulling attachment with teeth that clamp onto the post and lift it hydraulically.

Getting back to basics, you can make an A-frame post puller, as illustrated, by nailing together three pieces of wood. Wrap a cable around the base of the post, run it over the A-frame, and pull. The function of the frame is to convert horizontal pull into vertical pull, popping the post out as slick as you please. If muscle power doesn't do the trick, attach the cable to a vehicle or a come-along.

A word of caution that can't be repeated often enough: *Never* use a chain to apply horizontal pull. If the chain breaks, its recoil can injure or kill whomever it strikes.

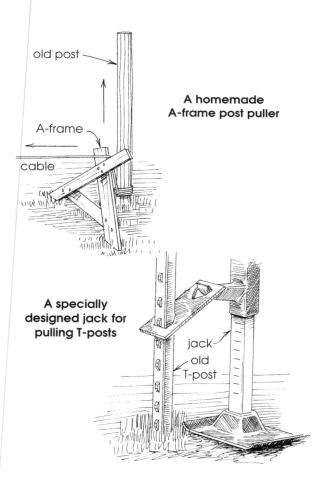

old post

A-frame

cable

**A homemade
A-frame post puller**

**A specially
designed jack for
pulling T-posts**

jack

old
T-post

FENCE WIRE

▾▲▾▲▾▲▾▲▾▲▾▲▾▲▾▲▾▲▾▲▾▲▾▲

The first known American to build fences out of wire was a Texan. The first smooth wire, then known as "snake wire," was patented in 1853. Barbed wire appeared three years later. Both kinds of wire have been constantly improved ever since.

PROPERTIES OF WIRE

Today, fence wire is selected on the basis of four properties: the material it is made of (usually steel), its tensile strength, how it is coated, and its thickness or gauge.

Tensile Strength

The tensile strength of wire measures how much it can be stretched before it breaks. The most common cause of wire fence failure is stretching the wire beyond its elastic limit. Overstretching doesn't necessarily occur only when the wire is initially strung. Horses or cattle may weigh it down to climb over, for example, or sheep, calves, and other small animals may spread wires apart to climb through.

Tensile strength is indicated as either "low" or "high." Low-tensile wire, occasionally called "merchant" wire and often called "soft" wire because it contains less carbon and is therefore physically softer, can be stretched only enough to eliminate sag. High-tensile wire has more built-in springiness and can be mechanically stretched until it is quite taut. Each manufacturer specifies how many pounds of force (Newtons, in metric) are needed to break their particular wire. High-tensile wire is made from plastic and aluminum alloy as well as steel. These alternatives were developed because high-tensile steel wire is stiff and can be difficult to work with.

Wire Coatings

Steel wire, the most common material used to make fence wire, is subject to atmospheric corrosion, causing the wire to take on the reddish brown hue of iron rust. Corrosion is the second most common cause of wire failure. It is hastened by humid air, acid rain, air pollutants, salt, chemical sprays, and moisture trapped by vegetation growing against the wire. Trellis wire is particularly susceptible to corrosion from chemical spray and moisture trapped by vines and leaves.

Quality steel wire is coated to protect it from exposure to air and other corrosive elements — at least until the coating is broken or itself becomes corroded with age. When you handle coated wire, take care not to drop the coil onto rocks or any other sharp objects that might nick the coating. In general, the cheaper the wire, the more poorly it is coated. The two most common metal coatings are aluminum and zinc.

Zinc-coated wire, called "galvanized," is designated as being Class 1, 2, or 3. The higher the number, the thicker the coating; the thicker the coating, the more resistant the wire is to corrosion. Most barbed wire is Class 1. So is most soft wire. High-tensile wire is Class 3, giving it more than twice the life of soft wire of equivalent gauge. Class 3 wire can be expected to last thirty-five to fifty years, varying with climatic conditions and damage to the coating. Avoid letting new galvanized wire of any class come into contact with old rusty wire; doing so causes an electrolytic effect that results in more rapid corrosion. The class number of galvanized wire is indicated on a tag attached to the roll of wire.

The amount of aluminum on aluminum-coated wire is seldom indicated on the tag. Aluminized wire costs 60 percent more than galvanized wire but is smoother, is more resistant to rust and corrosion, and lasts three to five times longer, especially in salty seaside air or where winter roads are cleared with salt. If the coating is broken and the wire is exposed to air, however, aluminum-coated wire doesn't hold up as well as galvanized — a significant consideration when you realize that whenever the wind blows, fence wire vibrates against its fasteners and wears away spots of coating. You can minimize the problem by using insulators to reduce friction between wire and fence posts, even if you're not putting up an electric fence.

Wire coated with aluminum-zinc alloy, sold under the trade name Galfan, lasts two to three times longer than zinc-coated wire but costs considerably less than aluminized wire. Another coating option is copper. Like aluminum, it resists weather better than zinc, but if the coating cracks it deteriorates more rapidly.

Yet another option is plastic-coated wire. Along with all-plastic wire, it is used primarily for horses (because its color makes it easier to see than metal wire) and in vineyards (because it isn't as sensitive to moisture as metal). Plastic and plastic-coated wire are less versatile than metal wire, since you can't later electrify a fence made with either. (A full discussion of wire suitable for electric fences appears in chapter 9.)

Wire Sizes

The weight of a roll of fence wire is determined not only by its length but also by the wire's diameter, indicated by its gauge number. Steel wire is sized according to the Standard Steel Gauge. The lower the gauge number, the larger the diameter of the wire. The thicker the wire, the stronger and more durable the fence will be. Also, the thicker the wire, the more it costs and the more difficult it is to handle — so there comes a point of diminishing returns.

In most cases, high-tensile wire has about twice the breaking strength of the same gauge soft wire. For comparable breaking strength, soft wire must therefore be thicker than equivalent high-tensile wire. For example, 13½-gauge high-tensile wire has approximately the same breaking strength as 12½-gauge soft wire.

STRINGING WIRE

How much a wire fence will cost depends on the quality of the wire you use and how many strands you string. The more strands you put up, the better your fence will work as a physical barrier. When in doubt, add more strands than you think you need.

The spacing of wire strands is indicated by a

LIFE EXPECTANCY OF GALVANIZED WIRE
Number of years before rust appears

Gauge	9	11	12½	14½
Dry Climate				
Class 1	15	11	11	7
Class 3	30	30	30	23
Humid Climate				
Class 1	8	6	6	5
Class 3	13	13	13	10
Salty/Polluted Air				
Class 1	3	2	2	1½
Class 3	6	6	6	4½
Years of useful life after rust appears				
Dry Climate	50+	50+	50+	50
Humid Climate	50+	50	35	20
Salty/Polluted Air	25	16	12	7

series of numbers separated with hyphens. A five-wire fence might have this designation: 12-10-10-8-8 (in metric: 30-25-25-20-20). The first number tells you how many inches (centimeters) the bottom wire is from the ground. The second number tells you how many inches (centimeters) the second wire is from the first, and so forth. To determine how far the top wire is from the ground, just add up

WIRE GAUGE

Number of Wire Gauge	Actual Wire Size	Steel Wire Gauge (U.S.)*	
		inches	mm
3	●	.244	6.190
6		.192	4.877
7		.177	4.496
8		.162	4.115
9		.148	3.767
10		.135	3.429
11	●	.121	3.061
12		.106	2.680
12½		.100	2.540
13		.092	2.324
13½		.086	2.184
14	●	.080	2.032
14½		.076	1.930
15		.072	1.829
15½		.067	1.702
16		.063	1.588
20	·	.035	.889

*also known as Standard Steel Gauge, Washburn and Moen, American Steel & Wire Co., or Roebling Wire Gage.

all the numbers. To calculate what length posts you will need for this fence, add together the height of the fence plus 6 inches (15 cm) plus the depth the posts will be set.

Procedure

Strands of wire strung parallel to each other along a fence line are called "line" wires. The general procedure for stringing line wire is to stretch one wire at a time between end posts. If any line post occurs on a rise or in a dip, attach the wire to the post as you pass it the first time. Otherwise, after secur-

ing the wire to the second end post, go back and loosely fasten it to intervening line posts. Then string the next wire.

To keep animals in, attach wire to the *insides* of line posts (the livestock side) so leaning critters can't push it free. To keep animals out (as for a garden), attach wire to the *outsides* of line posts. In either case, attach wire to the *outsides* of corner or curve posts so wire tension can't pull the wire loose. To reduce wear caused by rubbing at corner posts, hang a loose staple behind the wire as described in chapter 2, or use insulated tubing of the sort designed for electric fences, described in chapter 9.

Your fence will be stronger if, instead of running continuous line wire, you cut it at each anchor post, including corner posts, and start a new run. Rather than having one continuous fence, you will then have shorter runs coupled together at the anchor posts. You can, of course, make an exception where several anchor posts fall close together.

Besides giving you a stronger fence, starting a new run at each anchor post will help you maintain proper wire tension, as well as ensure that if one section is damaged (for example, by a fallen tree) the remaining sections will still be functional. Besides, repairing a damaged section is far easier and faster than restretching a whole fence.

Cutting Wire

Most stranded wire is rolled under tension. To keep it from recoiling and slapping you in the face or worse, secure both ends before cutting into the wire. If you are working with a partner, one of you can grasp the wire firmly, with hands about 12 inches (30 cm) apart, while the other cuts the wire. If you are working alone, grasp the wire with one hand, press it against the ground with a foot, and cut it between your hand and your foot.

To keep the loose end from recoiling when you let go, place it on the ground with a flat stone or heavy tool on top. Or push the end several inches into the soil. These precautions are especially important when you use high-tensile wire, which is stiffer than soft wire and has a stronger recoil when cut.

If any wire is left on a spool or coil, secure the end by poking it through a hole in the spool or by wrapping it twice around the coil. That way you can easily find the end next time; meanwhile the wire won't unwind and tangle.

TOOLS

Constructing a wire fence requires certain tools, some of which are fairly inexpensive. The costlier ones can be rented, but eventually you will want a full complement of your own tools for emergency repairs. You will find them in any well-stocked farm store or farm supply catalog.

One thing you will definitely need is a good pair of wire cutters. Standard sidecutters or pliers can be used for soft wire, but will quickly dull if used for high-tensile wire. For the latter you will need either a bolt cutter or a sidecutter designed specifically for high-tensile wire.

All-Purpose Tool

The all-purpose fencing tool, sometimes called an "all-in-one," is a multipurpose gadget that functions as a pair of pliers, a hammer, a staple puller, a wire grasper, and a wire cutter. Although it is highly overrated, it is relatively inexpensive and does come in handy.

Template

One tool you can easily make yourself is a template to help you accurately space wires against your posts. You could, of course, use a measuring tape, but a template is faster and reduces the chance of error.

A lightweight fiberglass rod works well for this, marked with paint stripes or an indelible marker to indicate the height of each wire. (For an electric fence, use red marks to indicate which wires are electrified and green marks for grounded wires.) Place some identifying mark at the top end of the template to avoid accidentally turning it upside down. If you are using a fiberglass rod that is pencil-pointed, make the pointed end the top. Otherwise, the point could dig into the ground, throwing your measurements off.

To use the template, rest it on the ground next to a post and, with the stripes as a guideline, mark off the wire positions. White chalk shows up best when marking dark-colored posts; a black felt-tip pen or a carpenter's pencil works fine on light-colored posts.

Wire Dispensers

All sorts of dispensers have been invented to simplify handling wire. The kind you use will depend on whether you are installing spooled or coiled wire. Barbed wire and smooth soft wire are usually spooled; high-tensile wire is usually coiled. No matter how it comes, you have to get the wire off the roll and laid out along the ground before attaching it to your posts.

A dispenser keeps the wire from looping or spiraling off the roll, which can make working with wire both difficult and dangerous, especially if it is

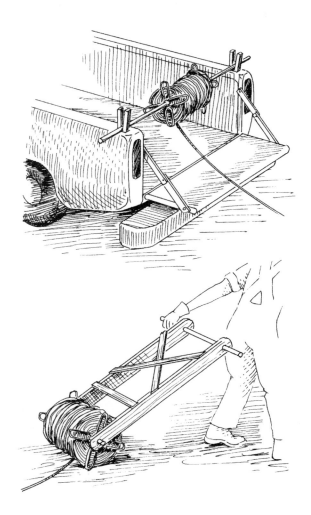

Homemade wire dispensers

barbed. Spiraling also causes kinks that weaken the wire and therefore weaken the fence. Even when you use a dispenser, guard against slack to avoid kinks. Since a kink reduces the breaking strength of wire by at least half, if you get a kink, cut it out and splice the wire back together. Splicing methods will be discussed later in this chapter.

The simplest dispenser for spooled wire consists of a steel post driven into the ground, around which the spool can revolve as you pull out the wire. Another dispenser for spools is a revolving handle that attaches to the center of the spool and lets the wire spin freely as you pay it out along the fence line.

More suitable for heavy spools is a longer handle that lets you pull the spool along the ground. Buy a handle made of metal, or fashion your own from wood. To make things easier, you might prefer to stick a sturdy rod through the spool and attach it to a wheelbarrow, pickup bed, or tractor hitch. You could, of course, carry the rod in your hands and let the wire unwind as you walk along.

The only way to avoid using a dispenser with coiled wire is to carry the coil and keep turning it as the wire pays out — a job that gets old fast. The alternative is to use a reel dispenser, called a "spinning jenny." Poke the reel's axle spike into the ground, then set the reel on top and let it rotate as you pull out the wire. If you use two reels with one stationed at each end of the fence run, you can save yourself a lot of extra steps by working back and forth from both sides. Some dispensers have a drag mechanism to prevent overspinning and tangles. If yours doesn't have such a mechanism, pull the wire more slowly as you come to the end of the run.

If your soil is too soft to hold the spiked version, or too hard to poke the spike in, use the footed version that sits on top of the ground, clamp the reel to the bumper of a pickup truck, or mount it on a tractor.

You might prefer a power dispenser for use with a pickup; it attaches to the bumper and is driven by the truck's battery. For a tractor, get a PTO-driven dispenser or even a multireel dispenser to speed things up. Be careful to stop the reels when you stop the tractor, otherwise they will keep paying out wire, causing all sorts of tangles and kinks.

Wire Stretchers

To tighten each line wire before securing it to the far anchor post, you will need a wire stretcher. Don't be tempted to use a moving tractor or pickup truck for stretching — it's far too easy to overstretch wire until it snaps, possibly injuring the tractor driver or some standerby. Even when you use a proper stretcher, wire can break and recoil if stretched too tightly. Wear hand and eye protection, keep children away, and try to position yourself behind a post while stretching is in progress.

The most common kind of stretcher is a rope pull stretcher. The most versatile is an ordinary come-along, also called a "cable jack." Both require an inexpensive wire grip to hook on one end. While you're at it, get a second grip for the other end, so you can later use the stretcher to splice broken line wire.

Two kinds of stretcher come with built-in grips. One, consisting of a row of teeth and a handle that ratchets along them, is designed primarily for barbed wire. Its serrated grips don't hold smooth wire very well. The other, called a "chain grab" stretcher,

A spinning jenny

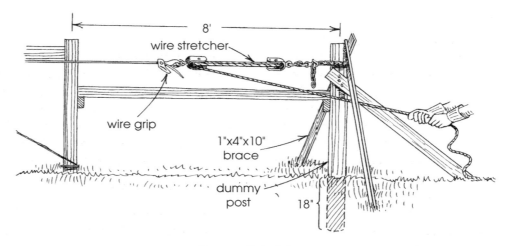

Rope pull stretcher

consists of a chain and a pair of legs that "walk" along the chain by grabbing on to alternate links. It is designed primarily for smooth wire and has a smooth grip designed to do less damage to the wire's coating than a serrated grip.

Although stretchers come in different styles, they all work basically the same way. You attach one end to something stationary, the other end to the wire, and work the stretcher until the wire is tight. Staple the wire firmly against the post. Slowly ease up on the stretcher until you are sure the wire won't let go, then finish wrapping and stapling the wire end.

In order to tighten wire sufficiently for a low-tension fence, the stretcher must be attached to a stationary object situated beyond the end of the fence run. That something could be a stout tree if one is in the right place, or a tractor or heavy truck. If nothing else is available, set up a dummy post to attach the stretcher to.

Place the dummy post in line with your fence posts, 8 feet (240 cm) beyond the second anchor post. Drive it at least 18 inches (45 cm) deep. Place a horizontal brace between the anchor post and the dummy post, and brace the dummy with three diagonal braces, as illustrated. All this may seem like extra work, but getting your fence wire properly tightened makes it well worth the effort.

A high-tension fence relies for tautness on in-line tighteners, discussed later in this chapter, making construction of a dummy post unnecessary. Instead, bring the end around the anchor post, attach it to one grip of the stretcher, and attach the other grip to the main wire. Crank until the wire is taut and fasten the end with an end splice (described later in this chapter).

No matter what kind of wire you are stringing, avoid stretching it tightly when the soil is wet; otherwise you could pull your anchor posts out of the ground. In summer weather, avoid stretching wire fully taut. Warm weather causes wire to expand and slacken, cold weather causes it to contract and tighten. If you stretch wire tight in warm weather, come winter it will contract and possibly snap. To avoid broken wires as a result of annual expansion and contraction, add compression springs or wire tighteners.

Compression Springs

Compression springs are used in wire fencing to give line wire flexibility to stretch in the event of heavy snow loads, extreme temperature changes, or pressure from rowdy livestock or unwary wildlife. They are generally used on very short runs, especially for a fence surrounding a holding pen where animals are crowded together. Compression springs can also be used to measure line wire tension, as described in the next chapter.

Suppliers sometimes mistakenly call compression springs "tension" springs. You should know the difference to make sure you are getting the right thing. A compression spring has two loops going down the center. When you pull on the ends, the spring's coils compress together. A tension spring, by contrast, has each end shaped into a loop or hook. When you pull, the coils move apart. The kind of spring that pulls your screen door shut is a tension spring. Compression springs, not tension springs, are used for fencing. They are fastened to the end of each line wire and attached directly to an anchor post.

Wire Tighteners

Wire tighteners are used to take up slack in line wire that sags as a result of age or overstretching. They are also used to maintain tension in a high-tension fence. Two main kinds of tightener are available: in-line tensioners and crimpers.

Crimpers, which look like a weird set of pliers, tighten wire by adding crimps or tension curves that create a wavy look, thus taking up slack by shortening the wire. The amount of sag you can take up per crimp varies with the crimper's design.

This tool works best for woven wire (see chapter 11) and old stranded wire that might snap if you bend it around an in-line tightener. On the other hand, very old wire that has lost its springiness may

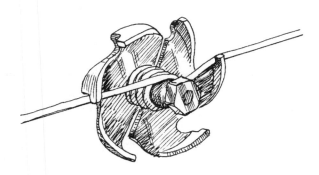

In-line tightener

not be capable of holding crimps at all. Wire strung under high tension won't hold crimps, either.

In-line tighteners, also called in-line "tensioners," "stretchers," or "strainers," "ratchet tighteners," or "tension reels," come in numerous styles. Some are suitable strictly for taking the sag out of low-tension fences. Others are designed for maintaining tension in a high-tension fence.

Regardless of the fence design, all in-line tighteners work essentially the same way: you attach one directly to the line wire and turn it to wind up slack. You thus need one or more tighteners for each line wire in your fence.

One design suitable for a low-tension fence consists of a heavy-gauge wire loop. You wind wire around the loop with the aid of an ordinary drill bit. Another is a metal disk with notches. When wire is wound against the disk, the notches keep it from unwinding.

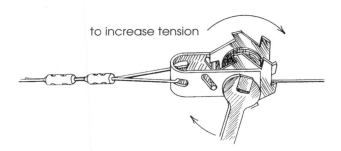

to increase tension

A wrench tightens the spool-type tightener.

A simple tightener used in Europe consists of a length of sturdy chain with one end fastened to the wire and the other end hooked on a nail in the anchor post. To tension the line wire, pull on the chain and slip the next link over the nail.

Adding any of these devices, however, reduces the breaking strength of high tensile wire. The most common tightener capable of handling high-tension consists of three parts: a ratcheted spool for the wire to wrap around, a brake pin that keeps the spool from unwinding, and a strap that holds the two together. Some spools can be turned with a screwdriver or pair of pliers, others with a socket wrench, and still others with a detachable handle purchased from the same source as the tighteners. Although you will need one tightener per wire, you need only one handle for the whole fence.

To install one of these tensioners, attach a wire stretcher to the line wire and pull out the slack. Cut the slack at midpoint, thread about 6 inches (15 cm) of one loose end through the holes in the tensioner's strap, and fasten the wire securely. Thread the other loose end through the hole in the spool, bend it around the spool, and cut off any excess. Use the handle or a wrench to turn the spool enough to hold the wire secure. Insert the brake pin (if it is not permanently affixed) and continue turning the spool until all the slack is taken up. Remove the wire stretcher.

An in-line tightener suitable for low- *or* high-tension fences consists of a spool you install without cutting the wire. However, if you use this device for a high-tension fence, initially you will have to string the line wires tighter than otherwise so the spools won't fill to capacity before the wires reach their proper tension. Use of in-line tighteners is discussed more fully in the next chapter.

Splicers

To connect the end of one roll of wire to the beginning of another, or to splice two ends of broken wire together, use one of several methods. To string soft wire or lightweight high-tensile wire, splice the ends with a figure eight knot. Standard high-tensile wire, on the other hand, is so stiff that knotting is difficult; besides, knotting both reduces wire's overall strength and breaks its galvanized coating.

A second soft-wire technique is the Western Union splice, which you can make with a pair of

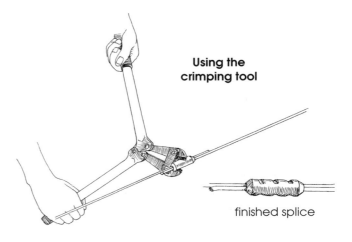

Using the crimping tool

finished splice

Compression sleeve

blunt-nose pliers or with a device known as a "wire splicer." Overlap the ends of wire at least 4 inches (10 cm), wrap each around the other at least five times, and cut off the excess. If you are splicing barbed wire, bring the two ends together, then separate the individual strands and wrap each one separately for a what is called a "split-strand Western Union splice." Like knotting, wrapping isn't strong enough for a high-tension fence.

You will get a stronger splice if you thread the wires through opposite ends of a little metal tube, then mash the tube to hold the wires tight. Such tubes are called "crimping sleeves," "compression sleeves," or simply "sleeves." The tool used to flatten them is called a "splicing tool," or sometimes a "crimping tool" or "crimper," even though it is nothing like the crimper used to take up slack in a sagging fence. This crimper is little more than a bolt cutter with the front part of its jaws blunted for crimping and the back part left sharp for cutting wire. Using one by yourself isn't nearly as easy as having a partner hold the wires and position the sleeves while you do the flattening.

Sleeves come in different diameters to accommodate different wire gauges. They also come in a longer and a shorter version. The standard short sleeve is for wires that overlap. The longer version is for butt splicing — joining, or butting, two wires without overlapping them. Sleeves sold under the brand name Nicopress are made specifically for high-tension fences.

Depending on who manufactures them, sleeves have two basic shapes. Just to keep everyone

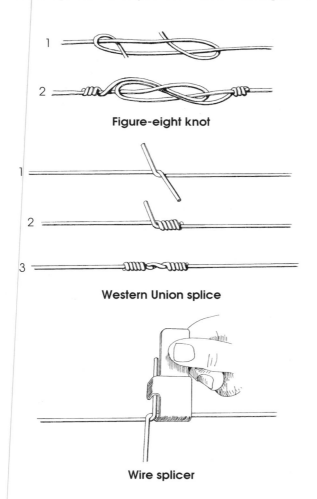

Figure-eight knot

Western Union splice

Wire splicer

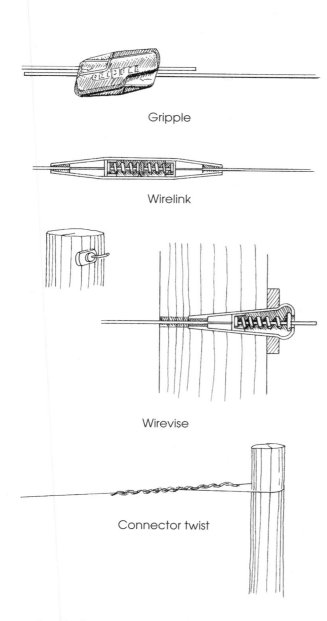

Gripple

Wirelink

Wirevise

Connector twist

bonded together and wound into a stiff spiral, this device is so simple it's almost unbelievable. All you do is butt the two ends of wire together and, using the center mark on the twist as a guide, wrap the connector around the wires so half goes around each wire. Connector twists come in various sizes to fit different wire gauges. Unlike sleeves, they can be reused.

Another strong butt-splice connector that doesn't require tools is sometimes called a "vise anchor" but usually goes by its trade name Wirelink. It consists of two sets of tapered springs inside a slender, cigar-shaped cartridge. Slip the wire ends into the ends of the cartridge, pull, and the springs grip them tightly. The tighter you pull, the firmer the springs grip. It works on the same principle as so-called Chinese handcuffs, those little woven tubes you get at carnivals — poke two fingers into the ends and you can pull all day without getting them back out. Wirelinks come in five sizes to fit wire from 5- to 16-gauge.

A similar device is the Gripple, consisting of a small metal rectangle with two parallel holes in it. Insert two wires into opposite ends and serrated rollers will lock them in place. Tension the wire by pulling on an end with a "Grippler" tool designed for the purpose. Gripples come in several sizes to accommodate smooth and barbed wire in different gauges.

As a general rule, use knots and wraps for temporary fences, splicing devices for permanent ones. Used for a high-tension fence, knots are effective only to 66 percent of the wire's breaking strength, butt-splice connectors to 80 percent. Connector twists, gripples, and properly crimped sleeves have 100 percent holding power.

End Splicers

To tie off wire at an anchor post, variously called "end," "dead-end," or "eye" splicing, you have basically the same choices as for splicing two ends together. For a low-tension fence, you can wrap the wire around the post twice, then wind it five times around itself. For lightweight high-tensile wire, tie a slip knot before winding the wire. Even with the knot, this end splice holds only to 60 percent of the wire's breaking strength.

A stronger method is to thread a compression

confused, the ones with an oval-shaped cross-section are called "tubular" and those shaped like a figure eight are called "oval." If you use the latter for a high-tension fence, you will need three, side by side. The crimping tool used to flatten oval sleeves has holes of different sizes ground into its jaws. Like the other version, this one doubles as a bolt cutter. Some brands will also crimp tension curves into sagging soft wire.

A strong splicing device that doesn't require any tools is the connector twist, also called a "tool-less connector," "vine line connector," "twist link tie," or "helical wire grip." Consisting of several wires sleeve onto the wire, wrap the wire once around the post, tuck the end into the sleeve, and crimp the sleeve shut. If you are using oval sleeves for a high-tension fence, you will need two, side by side. For fencers who can't remember to put the sleeves on ahead of time, "split" sleeves are available with slits along one side. Instead of threading them on the end, you slip them on sideways, but they won't hold as well as regular sleeves.

You can use a connector twist for end splicing. Bring the wire around the post, wrap half the twist around the end, and wrap the other half around the main wire. To use a Gripple as an end splice, slip it onto the wire, wrap the wire around the anchor post, and push the wire end into the second hole.

A modified version of the Wirelink, called a "Wirevise" or "link holder," is used for securing an end. It has one tapered spring enclosed in a cone-shaped cartridge. Drill a hole in the anchor post, parallel to the fence line. Slip the Wirevise in from the far side, insert the line wire through the other side, and pull with a pair of pliers until the wire is tight. Cut the wire end with 4 inches (10 cm) to spare, bend it downward, and staple it to the end post.

In a temperate climate where cold weather doesn't cause wire contraction, use a Wirevise or Gripple as a means of maintaining high tension on short runs of fence, no more than 200 feet (60 m) long. The problem in a seasonal climate is that, unlike an in-line tightener, a Wirevise can't be adjusted to release tension if line wires get too tight. Although Wirevises are fairly expensive, they are reusable, should the need arise.

LIGHTNING PROTECTION

An animal or person standing near a wire fence that's hit by lightning can be injured or killed — reason enough to stop working on your fence when lightning threatens. It is also a good reason to ground your fence properly. Even metal posts won't offer sufficient grounding if the soil is dry. Better to be safe and put in grounding rods. It is true they will make it more likely for lightning to strike your fence. But without them, a chance strike can travel for miles down the wire, causing serious damage. Grounding rods offer lightning strikes a fast path to the soil.

Ground a nonelectrified fence with ⅝-inch (16 mm) galvanized steel rods or 1-inch (25 mm) galvanized steel pipes, long enough to extend 6 inches (15 cm) above your posts and at least 5 feet (150 cm) into the ground or as deep as necessary to reach moist soil. Exactly how deep the rods must go depends on your climate and on your local ground water situation. The drier your soil, the deeper the rods must go.

If your fence wire is cut and wrapped at each anchor post, as it should be, you will need at least one grounding rod between each pair of anchor assemblies. In dry soil, place rods no less than 150 feet (45 m) from each anchor post and no more than twice that distance from the next nearest grounding rod. Where the soil remains moist year-round, you can double those distances. Don't worry as much about spacing rods evenly as putting them in low spots where the soil tends to stay moist.

Using a manual driver, drive the rods as close as possible to a post, on the same side as the wire. Secure the rod to the post so it makes firm contact, with the wire sandwiched between them. Lash the rod and post together with smooth wire or, if the post is wood, secure the rod with pipe straps. The procedure for grounding an electric fence is a bit different and is described in chapter 8.

Trees and buildings attract lightning. If you attach fence wire to either, you will increase the chance of a strike. Animals inside a barn to which fence wire is attached may be killed, or the barn may catch on fire. Instead of stapling wire to a building, affix it to an end post alongside the building. And *never* staple fence wire to trees — the practice is neither safe nor ecologically sound.

WIRE FENCES

▾▴▾▴▾▴▾▴▾▴▾▴▾▴▾▴▾▴▾▴▾▴▾▴

Strands of wire, strung parallel to each other to form a physical barrier, can be used to keep livestock in and some predators out. Smooth wire, either soft or high-tensile, can also be used to create trellises to confine grape and berry vines or dwarf fruit trees for easy maintenance and harvesting.

Barrier fences are made from three kinds of wire: barbed wire, cable, and high-tensile smooth wire. The latter is the most versatile because it can be adapted to nearly any situation. The super sturdiness of cable makes it suitable for corrals and barnyard lots where animals are closely confined. Barbed wire is best reserved for cattle on an open range; using it in confined areas or for active livestock is inhumane.

BARBED WIRE

A Texan was the first known person to use barbed wire when, in 1857, he topped his board fence with jagged metal strips to keep neighboring cattle away from his orchard. Word has it that his neighbors tried to run him out of town and, failing that, murdered him.

By the turn of the century, barbed wire had become a popular alternative to wood fences in the treeless prairie states, and some one thousand versions were developed. Barbed wire is still widely used for cattle and sheep in the Midwest. Although you can find only about half a dozen types today, nine different groups have dedicated themselves to collecting bygone variations.

"Barbed wire is good for two things," a friend is fond of saying, "tearing up pants and tearing up animals." That pretty well sums up the two primary purposes of barbed wire — to keep in cattle and to intimidate people. It works by pricking the skin and thereby deterring man or beast from going through. Barbed wire thus injures the very animals it is supposed to protect.

Because it is hazardous to humans, barbed wire is banned in most residential areas. Even where it is not restricted by law, it should not be used around horses, goats, hogs, or other active livestock, nor should it be used for registered stock, due to the increased likelihood of injury. The only reasonable use for barbed wire is to confine cattle.

Although a barbed fence is both cheaper and easier to maintain than a rail or woven wire fence, it costs considerably more than an electric fence. A three-wire electric fence, for example, is less expensive than a five-wire barbed fence, gives you better livestock control, and deters predators, which barbed wire does not.

Styles

Barbed wire comes on 80-rod spools. Since one rod equals 16.5 feet (5 m), one spool measures 1,320 feet (400 m), or exactly ¼ mile.

To the novice, barbed wire looks like barbed wire. An experienced eye, however, recognizes many subtle differences between one style and another. They all consist of two strands of smooth wire twisted together, and they all have evenly spaced barbs. The variations come in how thick the smooth wires are, how they're twisted, how thick the barbs are, how they're cut, how they're attached, and how far apart they're spaced.

If barbs were attached to plain smooth wire, they would slide back and forth unless welded in place — a highly expensive proposition. To hold the barbs, two strands of smooth wire are twisted together. Some brands have a right-hand twist, some have a left-hand twist, and some alternate after each barb. These variations result from different manufacturing processes and have little bearing on how the wire functions.

The twisted strands are usually either 12½-gauge (heavyweight) or 14-gauge (lightweight). The barbs attached to heavyweight wire may be either 12½-gauge or 14-gauge; to lightweight wire, 16-gauge. Barbs are made from short lengths of wire, cut at an angle to make a sharp point at each end. The exact shape of the points depends on the method used for cutting.

These short pieces of wire are wrapped at intervals along the twisted wire, the two ends left sticking out opposite each other to form two barbs. The wrap may be around only one strand of wire before the two are twisted together, or around both after they've been twisted. Sometimes two sets of barbs are wrapped at the same spot to make four barbs. Sets of barbs are spaced 4 or 5 inches apart (9 and 10 cm are standard in metric countries).

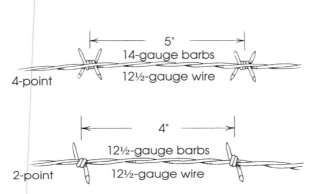

Two barbed wire styles

In selecting a style of barbed wire, cost is but one factor. Naturally enough, lightweight wire with two-point barbs is cheaper than heavyweight wire with four-point barbs. The former is used in dry climates and for large areas where animal pressure is low. The latter is best for humid areas and where animal pressure is greater. In a humid climate, you can expect 14-gauge barbed wire to last twenty years, while 12½-gauge will last about thirty-five years.

Barbed wire is protected with the same coatings as any steel wire: zinc, aluminum, or vinyl. The barbs of galvanized wire become dull as the wire rusts, while the barbs of aluminized wire stay sharp. Some barbed wire is made entirely of aluminum.

Two additional variations are worth noting. One is high-tensile barbed wire, which comes in higher gauges that are lighter yet stronger than the standard version. Finally, there is so-called "barbless" wire consisting of two strands of 12½-gauge wire twisted together without barbs. This variation is offered by barbed wire manufacturers for use around horses and for anchor post bracing.

Assembly

How well a barbed wire fence works is a function of wire tension, post spacing, and the number of strands. How many strands you need depends on the animals you wish to confine. For docile cattle, two strands may be enough, one at nose height and one at mid-chest — for cows, the spacing would be about 18-18 (45-45 cm), for calves about 12-18 (30-45 cm). For more active cattle, or for cows with calves, you will need four strands spaced 6-18-30-42 (15-45-75-105 cm). For bulls, use five strands spaced 12-10-10-10-10 (30-25-25-25-25 cm).

In a level, open field, space posts 16 to 20 feet (5 to 6 m) apart and add a spacer between each set of posts. On wide open range, you can space posts as far apart as 50 feet (15 m), with wire spacers every 10 feet (3 m) between them. For close confinement or rolling terrain, place posts 8 to 12 feet (2.5 to 4 m) apart.

On level terrain, single-span brace assemblies are sufficient for runs of 165 feet (50 m) or less. If a run is more than 165 feet (50 m) but less than 660 feet (200 m), you will need double-span assemblies. If the run is over 660 feet (200 m), divide it with pull-post assemblies. On hilly terrain, of course, you will need anchor assemblies at the top of each rise and at the bottom of each dip. Where the fence curves for more than 330 feet (100 m), add a pull-post assembly one span before and after the curve starts and ends.

Start stringing wire from the top down. Wind each wire twice around the first anchor post, then wrap it tightly around itself. Where the anchor post happens to be a gate post, remove barbs from the end before wrapping the post so you won't snag your clothing whenever you go through the gate. Exactly how you remove the barbs depends on whether they were added before or after the strands were twisted. If before, then separate the two strands and slip off the barbs. If after, pry the barbs off with pliers. If you are using wood anchor posts, wrap the wire before stapling it in place.

Unwind the wire along the inside of the fence line, going a bit beyond the last anchor post. Then stretch the wire with a stretcher or come-along, first untwisting the strands so they don't bite into each other. Cut one of the strands about 2 feet (60 cm) beyond the anchor post, keeping tension on the remaining strand. Remove the barbs, wind the loose strand around the anchor post, then wrap it around the line wire, leaving enough space between each turn to interwrap the second strand.

Cut the second strand and wrap it around the post, then around the wire. To string high-tensile barbed wire, wrap the end three times around the post and at least five times around itself. Now go along the fence line and attach the wire to T-posts with clips or to wood posts with staples. Remember to allow enough space for the wire to move freely behind the staples as it expands and contracts with the weather.

Because barbed wire consists of two strands twisted together, it has a bit of springiness to it. It is therefore not easy to tell when the wire is properly stretched. Many fencers pull it too tight, causing it

to break and possibly injure someone. Even if it doesn't break while you're working, overstretched low-tension barbed wire is likely to snap when it shrinks in cold weather or when pressure is put on it by animals or machinery — the main reason barbed wire fences require constant repair.

A good rule of thumb is that free-hanging barbed wire should sag no more than 12 inches (30 cm) at midpoint when stretched 130 to 160 feet (40 to 50 m). Check this before securing the wire to line posts by threading it through loose smooth-wire loops attached to appropriately spaced line posts, then measuring how much it sags between the loops.

CABLE FENCES

Cable fences are used to create corrals, barn lots, holding pens, and other areas where horses or cattle are held in close confinement. Cable, also called "wire rope" or "cattle strand," is fairly expensive but makes an extremely strong fence. Standard fencing cable consists of several strands of thin, smooth wire twisted together to a diameter of ⅜ inch (1 cm). These twisted strands readily trap

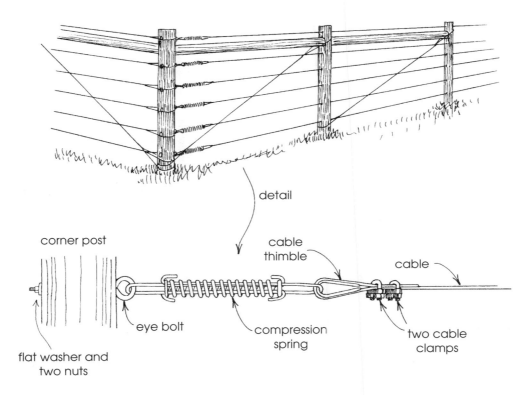

detail

corner post

cable
thimble

cable

eye bolt

compression
spring

two cable
clamps

flat washer and
two nuts

Cable fence

moisture, causing the cable to rust rather easily.

The number of cables, and their spacing, varies with the size of the animal being confined. A typical fence might be 50 inches (136 cm) high and have six cables spaced 12-8-8-8-7-7 (30-20-20-20-23-23 cm). Line posts would be 5 inches (13 cm) in diameter, 8 feet (240 cm) long, 8 feet (240 cm) apart, and set 3 feet (90 cm) deep. The cables are threaded through ½-inch (13 mm) holes in the posts.

Wood corner posts are 6 inches (15 cm) in diameter, 9 feet (270 cm) long, set 4 feet (120 cm) deep, and supported by double H-brace assemblies. One end of each cable is wrapped tightly around a corner post. The other end is fastened to a heavy compression spring, as illustrated. The spring is hooked into a 7-inch (178 mm) eyebolt, which is threaded into a ⁹⁄₁₆-inch (14 cm) hole in the corner post and held with a flat washer and two nuts.

To connect each cable, thread the cable through the appropriate holes in the line posts. Slip a ⅜-inch (1 cm) cable thimble through the end of the spring, loop the cable around the thimble, and clamp it down with two ⅜-inch (1 cm) cable clamps, nuts pressed against the full-length side (not the cut end) of the cable. Cable thimbles and clamps are available at any good hardware store.

Pull the cable taut at the other end with a wire stretcher. Wrap the end around the corner post and secure it with two more cable clamps. Adjust cable tension by loosening the eyebolt nuts — under proper tension, the compression spring should just start to open.

Tension Version

Thanks to high-tensile wire, you can construct an equally strong but less expensive corral by threading closely spaced strands of high-tensile wire through line posts, attaching one end with a Wirevise and the other with a compression spring and eyebolt. To increase tension, pull the wire through the Wirevise. To relieve tension, unscrew the eyebolt nuts a turn or two. A typical barnyard lot for cattle has ten wires, spaced 10-4-4-4-5-5-5-5-5-5 (25-10-10-10-13-13-13-13-13-13 cm).

HIGH-TENSION FENCES

All high-tension fences are constructed of high-tensile wire, but not all fences made of high-tensile wire are high-tension fences. "High-tensile" refers to the wire's strength. "High-tension" refers to its tautness, which should be great enough so that the wires can't be easily pried apart and that they bounce back on sudden impact.

Owners of high-tension fences love to trade stories about line wires that spring back into place after being pinned to the ground by a fallen tree, flattened by a rolling bale of hay, or challenged by a bull elk. In many such cases, staples pop loose and posts bend or break, but the line wires remain taut, keeping the fence fully functional.

In similar situations, soft wire, instead of bouncing back, stretches and breaks, so is unsuitable for a high-tension fence. High-tensile wire, on the other hand, doesn't *have* to be stretched under tension. It can be, and often is, used for low-tension fences, usually electrified. The following description relates only to nonelectrified high-tension fences; electric fences are discussed in chapter 9.

Pros and Cons

High-tension fences have been around for thirty-five years or more, so we know that a well-built one will last at least that long. The concept originated in Australia and New Zealand and was first used in the U.S. in 1973. The style is variously known as "New Zealand fence," "high-tension fence," or simply "tension fence."

Its advantages are its great strength and durability coupled with its relatively low cost per year of useful life. A properly constructed tension fence will outlast a barbed wire or woven wire fence and requires much less maintenance, consisting primarily of keeping back weeds.

While tension fences are extremely versatile, they are by no means suitable for all situations. Installing one on hilly terrain or where the fence line zigs and zags, for example, would be very expensive because you would need so many extra posts. A tension fence isn't suitable, either, for confining small animals under crowded conditions, since they might push each other through. Even a big animal like an elk will charge through a tension fence, if sufficiently motivated. Finally, you certainly wouldn't put one up if you weren't sure the fence was permanent — constructing a tension fence entails a lot of work and removing one isn't easy.

Nonelectrified tension fences *are* suitable as permanent boundary fences and interior cross fences on nearly level terrain where long, straight runs are possible. They work best with large livestock such as horses, cattle, and big game animals, but have also been used successfully to confine sheep, pigs, and goats. While tension fences aren't picturesque, they are tidy and attractive, and therefore less likely than other kinds of farm fence to cause the disapproval of neighbors.

Installation

The key to an effective tension fence is high-tensile wire able to withstand not only constant fence tension, but also any increase in tension as a result of animal impact or cold-weather contraction. High-tensile wire can be strung four times tighter than low-tensile wire and therefore doesn't have the same tendency to sag in warm weather. It also rapidly returns to uniform tension after impact has stretched it to as much as 75 percent of its breaking point. A standard nonelectrified tension fence has eight to ten strands, closely spaced.

High-tensile wire as thin as 16-gauge, sometimes called "spider wire," is occasionally used for cross-fencing, although it's not nearly as convenient as the temporary fences described in chapter 10. Commonly used for permanent perimeter fences are 14½-gauge wire, which is easy to handle and easy to knot, and 12½-gauge, which is considerably stiffer but three times stronger than 16-gauge. One coil contains 4,000 feet (1,215 m).

Proper tensioning of the wire requires super strong, double H-brace anchor assemblies with 6-inch-diameter (15 cm) anchor posts, 8 feet (240 cm) long, set 4 feet (120 cm) deep. Lesser anchor posts would be lifted out of the ground by fully tensioned line wires. For a very long fence line, include a pull-post assembly every ½ mile (800 m).

Line wire spacing is achieved by means of posts and spacers set close enough together to accommodate back scratching, pushing, and other typical livestock activities. Line posts can be of wood, steel, or concrete; spacers can be of wood or fiberglass. How far apart you set the posts and spacers depends on the nature of the terrain and how crowded your animals are. For horses, posts are often spaced closer together than otherwise. This makes the fence more visible, thereby decreasing injuries.

The closer posts are spaced, however, the greater the load imposed on them during impact, and the greater the chance that they will overturn, bend, or break. In a tension fence, it is the wire, not the posts, that serves as a physical barrier. On open, level range, posts can be set as far apart as 100 feet (30 m) with spacers every 33 feet (10 m) between them.

For a typical pasture on level terrain, space the posts 30 feet (9 m) apart with spacers every 10 feet (3 m). If the pasture will be heavily grazed, place spacers every 5 feet (1.5 m); if lightly grazed, you can get by with one every 15 feet (4.5 m). For pasture on hilly terrain, more closely spaced posts will be needed, as dictated by the locations of dips and rises.

Closer spacing is also necessary if you are building a barnyard pen or corral where animal pressure is fairly high. In such a case, place posts 24 feet (730 cm) apart with spacers every 8 feet (240 cm). For very close confinement, set the posts 8 to 12 feet (240 to 360 cm) apart and either thread line wires through holes drilled in the posts (as for a cable fence) or weave the wires in and out between the posts before fastening them on.

On open, level terrain, build a suspension fence by spacing posts 65 feet (20 m) apart and using spacers that don't touch the ground. The resulting fence will give under stock pressure and spring back when the pressure is removed. The springiness also discourages stock from rubbing on the fence or trying to get through.

The general procedure for installing a tension fence is: construct anchor assemblies, string the bottom line wire as a guide, set the line posts, string the rest of the line wires, install in-line tensioners (described below), fasten the wires to the line posts, put in the spacers. If the ground is uneven, attach each wire to the rise and dip posts *before* installing the tensioners. One day after putting up the fence, retighten the in-line tensioners a click or two.

In-Line Tensioners

Each wire of a tension fence must be tightened with one or more in-line tensioners, installed after all line wires have been initially stretched and fastened to the anchor posts. Exactly where along each wire you place the in-line tensioner depends on the length of the fence line and whether the terrain is level or sloped.

If the fence goes up a hill, install tensioners near the downslope anchor post. If the run is less than 600 feet (180 m), put the tensioners near an anchor post or gate post for easy access. If the run is more than 600 feet (180 m), install tensioners at midpoint for even tensioning.

Exactly how much wire you can tighten with one tensioner varies with the brand. A typical tensioner can handle a straight run of up to 3,000 feet (915 m). If the distance between anchor posts is more than that, add two tensioners to each wire at evenly spaced intervals.

Since corner and dip posts increase friction, they decrease the length of wire one tensioner can handle by about 500 feet (150 m). Add a second tensioner if the line wire goes around two 90° corners.

Install tensioners at least 4 feet (120 cm) from the nearest post so you have room to crank the handle. Since a tensioner creeps forward as it takes up slack, orient the spool side away from the post so the tensioner won't move toward it.

After all tensioners are installed, tighten the wires one by one, starting at the top to give anchor posts a chance to settle. If you start at the bottom, the anchor posts will pull inward as you work your way up, and you'll have to tension all the wires again.

Once your fence is in and properly tensioned, you will need to adjust the tension only infrequently if you live in a temperate area. If, however, the annual temperature range in your area exceeds 100°F (56°C), you will need to reduce the tension each fall and increase it again in spring. As a general rule, for every 10°F (5°C) above or below 60°F (15°C), subtract or add 10 pounds (50 N) of tension on standard 12½-gauge wire.

Measuring Tension

For the fence to function properly, its line wires must be kept under just the right amount of tension. If the tension is too low, livestock can push the wires apart and get out. If tension is too high, anchor posts will pull out and brace assemblies will buckle. In addition, overtensioned wire can break and recoil, causing serious injury to eyes or other parts. Wearing gloves and safety goggles isn't a bad idea when you work with high-tensile wire.

Even if the wire doesn't break, overtensioning can stretch it beyond its elastic limit or "yield point," causing it to remain permanently stretched. As a general rule, stretch high-tensile wire only to about one-third its minimum breaking strength to ensure that additional stretching due to impact or high temperature won't push it beyond its yield point.

In New Zealand, where tension fences originated, the tension on 12½-gauge wire is initially set at 440 pounds (1,980 N), settling to 200 pounds (900 N) during the first year and 150 pounds (670 N) thereafter. Here in the U.S., we use permanent in-line tensioners, which lets us start with a somewhat lower initial tension — generally 300 pounds (1,340 N), later adjusted to 250 pounds (1,130 N).

Some people don't bother to measure tension exactly, especially when stringing 16-gauge wire. They just stretch the line wires enough to take out sag and retension periodically as needed.

Measuring tension isn't difficult, and doing so ensures that the fence will function the way it is supposed to. If your in-line tensioners are the sort that can be tightened with a socket wrench, measure tension directly with a torque wrench. A torque of 12 foot-pounds (5.5 kg) is approximately equal to 250 pounds (1,130 N). If you have a chain-grab wire stretcher, fit it with a tension indicator available from the same supplier.

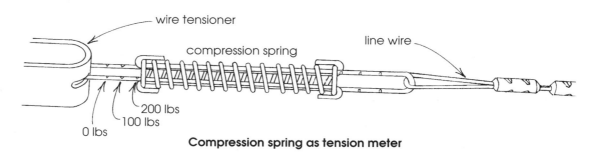

Compression spring as tension meter

Perhaps the most common way to measure tension is with a compression spring installed next to the in-line tensioner. When the spring is shortened 1½ to 2 inches (38 to 50 mm), depending on the brand, the wire is under 250 pounds (1,130 N) of tension. Some brands make measuring easy — they have tension indicator marks, each notch representing approximately 100 pounds (450 N).

In a tension fence, compression springs should be used only to indicate tension, not to keep the wire taut or cushion it against impact. (Two exceptions are where a run measures less than 150 feet (45 m) or where animal pressure is high — such as in a corral or barnyard pen.) You can therefore save money by placing a spring on one line wire only, then tensioning the rest by feel and sight. Pull each wire toward you the same distance and judge whether it offers the same amount of resistance as the wire tensioned with the spring.

Homemade Tension Meter

Another alternative for measuring tension is a simple tension meter you can make with a spring scale (of the sort used to weigh fish), three nails, and a piece of lumber at least 42 inches (110 cm) long. Drive two of the nails into the board, 40 inches (1 m) apart. Draw a straight line from one nail to the other and drive in the third nail halfway between them and ½ inch (1 cm) below the line.

When you use this device, measure tension halfway between the in-line tensioner and the farthest anchor post. Rest the line wire against the outer two nails. At the midpoint between the nails, hook the spring scale over the wire and pull the wire toward you, parallel to the ground, until the wire touches the middle nail. Multiply the number of pounds indicated on the scale by 20 (multiply by 200 for kg). If, for example, the scale reads 7½ pounds (3.4 kg),

the wire is under approximately 150 pounds (670 N) of tension. When the scale reads 12½ pounds (5.6 kg), the wire is under a full 250 pounds (1,130 N).

Line Wire Spacing

How many strands of wire you string and how far apart you space them depend on what kind of animals you plan to confine, their size, how active they are, and how crowded they are. Large, lethargic animals in uncrowded conditions can be confined with fewer line wires spaced farther apart. You will need more wires spaced closer together for smaller, more active, and more closely confined animals that could slip, or be pushed, through.

If you'll be confining animals of differing natures, say horses and foals or cows and sheep, design your fence for the smallest and/or most active. Since the bottom one-third of any fence experiences the most animal pressure, line wires are typically spaced closer together toward the bottom and progressively farther apart as the fence reaches optimum height.

To keep in hard-to-confine animals or to keep out persistent predators, augment your tension fence with electrified line wires or offset wires (described in chapter 9). Electrification serves the additional purpose of protecting the fence from back rubbing, climbing, and pushing toward greener pastures. Typically, the second and fourth wire from the bottom are electrified to discourage digging and crawling, the top wire is electrified to discourage climbing, and the wire closest to nose-height of the animal is electrified to discourage pushing while grazing. The next few chapters describe electrified fences in detail.

TRELLISES

Constructing a vertical trellis is much the same as building a wire barrier fence. The main difference is that you need fewer line wires spaced farther apart. The trellises described here are suitable for backyard plantings of grapes, raspberries, blackberries, and espaliered dwarf fruit trees. If you are interested in trellises strong enough for standard fruit trees or designed for mechanical pruning and picking, consult *How to Build Orchard and Vineyard Trellises,* listed in the appendix. Temporary trellises

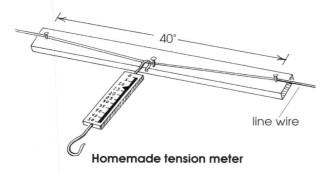

40"

line wire

Homemade tension meter

NONELECTRIC TENSION FENCE WIRE SPACING

Livestock	Wires	Spacing (in.)	Spacing (cm.)
Cattle, horses	8	4-5-5-5-6-6-7-8	10-13-13-13-15-15-18-20
Pigs	10	2-2-2-2-2-4-4-4-6-6	5-5-5-5-5-10-10-10-15-15
Sheep, goats	10	4-4-4-4-5-5-5-5-5-5	10-10-10-10-13-13-13-13-13-13
Foals	12	6-4-4-4-5-5-5-5-5-5-5-5	15-10-10-10-13-13-13-13-13-13-13-13

for such seasonal garden plants as tomatoes, peas, or beans are described in chapter 10.

Although you can put in a permanent trellis any time, building one before you plant is faster and easier. The specific design you choose will depend on the kind of plant you grow and the training system you use. You'll have to seek elsewhere for a discussion on the relative merits of various training systems, a subject that goes beyond the scope of this book.

All trellises share certain characteristics. They have strong, well-anchored end assemblies, evenly spaced line posts, and one or more tightly strung line wires that are grounded, like any wire fence, for the protection of plants as well as humans. To take the best advantage of available sunlight, orient your trellis north to south.

Trellis Posts

A standard wood trellis post, sometimes called a "vineyard" post, is 8 feet (240 cm) long and 3 to 4 inches (76 to 100 mm) in diameter. The bottom may be either blunt cut or pencil-pointed for driving. Sometimes sawn 2x2 (5x5 cm) "grape" stakes are used, but they have one-sixth the strength of a standard post and therefore must be spaced much closer together, usually at the rate of one per plant. Standard vineyard posts can be spaced every three or four plants, or even farther apart. Standard end posts are 8 to 10 feet (240 to 300 cm) long and 4 to 7 inches (100 to 178 mm) in diameter.

Wood posts should be pressure treated, since vines trap moisture against posts and hasten decay. Decaying posts attract insects to plants and allow

line wires to sag. The most widely used preservative for trellis posts is CCA, which isn't readily absorbed by plants and won't burn them. Penta- and creosote-treated posts can chemically burn plants for the first few years. You can avoid these problems by using concrete or steel posts instead of wood.

Sink end posts 3 feet (90 cm) deep, leaning slightly away from center. Set line posts at least 2 feet (60 cm) deep. If you live in a hard-frost area, drive them deep enough to get below the frost line and prevent heaving.

If, instead of driving, you set posts in drilled holes, use the same auger to make planting holes. Unlike fence posts, which are set perpendicular to the ground, trellis posts are set plumb. On uneven terrain, therefore, the height of the trellis will vary in relation to the ground, perhaps requiring some line posts that are longer (or shorter) than the rest.

Trellis end posts have traditionally been braced with guy wires, a practice that takes up valuable space, makes weed control difficult, and guarantees you will trip over a guy more often than you'd like. The ideal brace assembly for a trellis allows you to train plants right to the end post, with no wasted space.

One alternative, suitable for low-tension trellises in soil that's neither loose nor wet, is to brace wood end posts with a deadman. For a high-tension trellis, the best alternative is H-brace assemblies. Single-span assemblies are adequate for most backyard trellises.

Attach line wires to the windward side of posts so blowing foliage will push the wires against the posts, rather than pulling them away. If you need extra height, staple the top wire to the tops of posts.

The kind of plant you grow and the training system you use will determine how many wires you string, what gauge you need, and whether a low-tension or a high-tension trellis will work better.

Conventional vs. High-Tension

Maintaining a trellis primarily involves repairing broken wires, replacing rusty ones, and retensioning wires that have sagged because of weather changes or heavy fruit load. A high-tension trellis won't rust as rapidly as a conventional one and can better withstand the extra tension caused by wind, low-temperature contraction, and heavy fruit loads; therefore it requires less maintenance.

Conventional 12-gauge soft wire can be tensioned to only 100 pounds (445 N), can bear a load of 700 pounds (3,115 N), and is Class 1 galvanized. High-tensile 12½-gauge wire can be tensioned to 300 pounds (1,335 N), will handle a load of 1,200 pounds (5,340 N), and is Class 3 galvanized so it lasts more than twice as long. A tension trellis will outlive most plants and should last long enough for a second or even a third planting.

With high-tensile wire you can space line posts as far apart as 30 feet (915 cm), depending on wire tension, crop load, wind, and terrain. For a low-tension trellis, you need posts every 18 to 21 feet (550 to 640 cm), or roughly between every three or four vines. The money you save on posts, however, will be spent on in-line tensioners.

For a trellis over 3,000 feet (90 m) long, you will need two tensioners per wire, each one-quarter of the distance from an end. For a trellis spanning less than 3,000 feet (90 m) but over 500 feet (150 m), install one tensioner halfway along each wire. If the trellis is less than 500 feet (150 m), you can put the tensioners near either end.

If you live in a temperate climate where wire contraction during winter isn't a problem, and your trellis is no more than 200 feet (60 m) long, you can get by using Gripple or Wirevise end splicers to maintain tension. If there is any chance your line wires will get too taut, install eyebolts at the other end of each wire, as for a barnyard corral.

Berry Trellises

For erect-growing berries, you don't really need a trellis. But using one helps keep canes within a confined area, simplifies mowing and weeding, and lets you see and prune damaged canes more easily. For raspberries, two wires are commonly used, spaced 36-66 (90-165 cm). For biennial varieties, tie the eight strongest canes to the wires and cut off the rest at ground level. In late winter, prune each cane to a leaf bud just above the top wire. The following summer after harvest, cut back the old canes and tie

AVERAGE SPACING FOR TRELLISED PLANTS

Plant	Between Plants		Between Rows	
Blackberries				
erect	5 ft.	150 cm	8 ft.	245 cm
trailing	5 ft.	150 cm	8 ft.	245 cm
trailing thornless	8 ft.	245 cm	10 ft.	305 cm
Raspberries	18 in.	45 cm	6 ft.	180 cm
Grapes				
European	6 ft.	180 cm	9 ft.	275 cm
American	8 ft.	245 cm	9 ft.	275 cm
Muscadine	20 ft.	610 cm	9 ft.	275 cm
Fruit Trees				
dwarf	8 ft.	245 cm	12 ft.	365 cm
semi-dwarf	10 ft.	305 cm	15 ft.	460 cm

up the eight strongest new ones. For ever-bearing varieties, each winter either prune out weak canes and cut back old ones to live wood or, for a single crop of larger berries, cut all canes at ground level and train new growth to the trellis.

For erect varieties of blackberry, tie canes to a single wire, strung 40 inches (1 m) above ground. If you prune away side shoots, you will lose a little fruit but the planting will be much easier to manage.

For trailing blackberries, a trellis keeps fruit clean and off the ground and makes picking easier. The traditional method of training canes to two wires, spaced 48-18 (120-45 cm), letting new canes trail along the ground below the trellis. A more sensible approach is a three-wire trellis, spaced 24-12-12 (60-30-30 cm). Tie old canes in a fan shape; tie new growth to the bottom wire.

Grape Trellises

Before building a trellis for grapes, decide which training system to use. For the single-curtain system, which calls for pruning to two cordons (fruiting canes) and letting shoots hang down, you will need only one wire. It can be 5 to 6 feet (150 to 180 cm) above ground, depending on what is a comfortable working height for you.

The vertical hedgerow system, also called the "Guyot" or "two-cane Kniffen" system, requires similar pruning but calls for shoots to be trained upward and woven between line wires. Here you will need four wires, spaced 24-12-12-12 or 30-14-14-14 (60-30-30-30 or 76-35-35-35 cm). Again, choose a comfortable working height.

The double-curtain or four-cane Kniffen system calls for four cordons with shoots hanging down and is used primarily for vigorous American cultivars such as Concord and its hybrids. Cordons are trained along two wires, spaced anywhere from 30-30 to 36-36 (76-76 to 90-90 cm). This system is not as popular as the others because fruit on the lower canes gets shaded by the upper canes.

Grapevine and Tree Trellises

The most versatile trellis is designed for the "leaf palmette" system of training dwarf and semi-dwarf fruit trees as well as grapevines. Between two and five wires may be strung; two are most common for grapes and three for espaliered fruit trees.

Typical spacing for a two-wire trellis is 30-28 (76-71 cm). For a three-wire trellis it is 30-20-20 (76-51-51 cm). If you're building a three-wire trellis for fruit trees, be on the safe side and brace end posts with double assemblies.

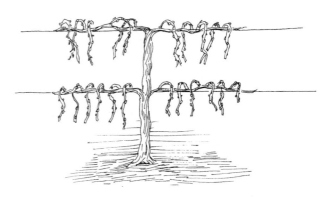

Double-curtain trellis

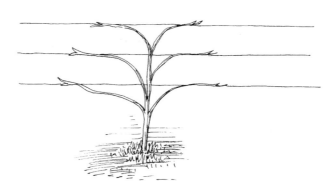

Leaf palmette trellis

CHAPTER 7

SELECTING AN ENERGIZER

▾▴▾▴▾▴▾▴▾▴▾▴▾▴▾▴▾▴▾▴▾▴▾▴

There are two basic types of agricultural fence: physical barriers and psychological barriers. Physical barriers are constructed so animals can't get through; psychological barriers are designed to intimidate animals into not *wanting* to get through.

For most animals, the threat of a strong electric shock is sufficiently intimidating. Just how much deterrent value an electric fence has depends on its conductivity (the subject of chapter 9) and on the effectiveness of its energizer (the subject of this chapter).

PROS AND CONS

An electrified fence is the cheapest kind of fence, the easiest to construct, and potentially the longest lasting since animals don't rub against it, climb on it, or push through it. A properly built electric fence offers greater protection from predators and more effective control of livestock than any other fence.

An electric fence does, however, have certain disadvantages. Unless it is also built to function as a physical barrier, it must be electrified at all times to be effective. This means you will have to patrol the fence periodically, eliminating shorts. If you use battery power, you will have the additional concern of making sure the battery is sufficiently charged.

Another disadvantage is that electrical conductivity varies with climactic conditions. Unless you design your fence with this in mind, it will be less effective in dry or snowy weather and will short out easily in wet weather.

You may also have to alter your management practices—no longer can you crowd animals against a fence corner for routine handling. You will have to herd them into a barn or corral.

An electric fence is not something you can throw together. In some countries, the law requires you to hire a licensed electrician to put one up. In the U.S., it is still legal to install your own fences, but with that freedom comes the obligation to use common sense. For starters, never make your own energizer. You could end up causing a fire or, worse, injuring or killing someone. The few dollars one fellow saved by converting an appliance transformer cost him the life of his five-year-old child. Another child died after falling on an electrified wire strung around a neighbor's garden and plugged directly into a wall socket. Accidents of this sort don't happen with a properly constructed electric fence.

OLD VS. NEW

Energizer, fence charger, fencer, controller, power unit—all these terms refer to that little box responsible for the big jolt. It works by converting household electricity or battery-generated power into higher voltage, then sending the charge through fence wires in a series of short bursts. You can actually hear a slow, steady click-click-click as the pulses go out.

The first energizer invented in the U.S. was patented around 1893. The idea wasn't developed commercially until the early 1930s, when two different companies began independently making battery-operated controllers that sold like hotcakes during the '40s and '50s.

The first plug-in model, a "weed burner," came along in the mid-50s. The long, slow pulse of those early controllers dissipated rapidly, so only short fences could be energized. Even short fences easily became ineffective because of energy leaks through faulty insulators and encroaching vegetation.

Early controllers were therefore beefed up to put out more energy. The problem was, the pulse was on so long — sometimes as much as half a second — that weeds touching the wire could heat up and catch fire. Wires could melt. Animals or people getting tangled in a fence could be electrocuted.

To reduce these dangers, resistors were added to impede or restrict the flow of energy, hence the designation "high impedance" for this kind of energizer. But restricting the energy flow caused the current to dissipate gradually as it pulsed down the wire, rendering the fence ineffective. Back to square one.

By the early 1960s, it had been discovered that a fence is less likely to short out if you send a large amount of energy through the wires in a very short burst. Because this new kind of controller puts out such a short pulse, it is called a "short shock" system. It is also called a "high-energy, low-impedance" system because it puts out a lot of energy without restricting energy flow. The sharp, short pulses travel quickly through line wires and remain effective for many miles.

Short-shock energizers are three times more powerful than the old controllers, yet are extremely safe. Despite their high voltage, the pulses last as little as 0.0001 seconds and are spaced about 1 second apart. Touch the fence and you will get quite a shock, but the ensuing off-time lets you pull back and recover.

MAKING YOUR SELECTION

Fence controllers come in various sizes, referring not to a unit's bulk but to the amount of energy it puts out. The right energizer for your particular fence creates an effective sphere of intimidation around each pulsed wire, discourages vegetative growth, and overcomes minor fencing flaws. Trying to save money with an energizer that's too small can render your entire fence useless. Getting one that's too big is not only a waste of money, but can cause sparking or arcing across insulators.

The proper size for you will depend on the length of your fence, how many pulsed strands it contains, what wire you use, what animals you want to contain, and how much weed encroachment is likely — especially during seasons of rapid growth when energy drain is greatest. Factor in extra power if, like most folks, you don't diligently spray or mow

along the fence line. Allow more power yet to confine hard-to-control animals that require more than the usual number of pulsed strands.

Energizer quality varies widely and manufacturers' claims — often based on creative calculations — cannot easily be verified. A good energizer isn't cheap. Neither are some bad ones, making it important to be an informed buyer. Unfortunately, the electric fence industry hasn't developed a system by which one brand can be compared to another. The American Society of Agricultural Engineers has been trying to establish common standards, but so far compliance is strictly voluntary.

Since new units are constantly coming out, your best bet is to find an experienced electric fence user or a knowledgeable supplier whose brains you can pick. Some conscientious suppliers will freely discuss the pros and cons of each brand, including their own. Don't buy an energizer from anyone — including the clerk at your local farm store — who can't explain its output to your satisfaction.

Measuring Output

Various manufacturers have different ways of measuring output. To understand what they are talking about, you will have to familiarize yourself with certain terminology.

"Miles of fence" is one way to measure an energizer's output. The term actually refers to miles of line wire, since output is measured in a single continuous strand, 3 feet (90 cm) off the ground and free of weeds — in other words, ideal laboratory conditions. The distance a pulse can travel in actual field conditions is considerably less because of climate, weed load, fence construction, wire gauge, and the electromagnetic resistance arising when you string two or more wires parallel to each other.

The output of most energizers is also rated in volts and amps or joules. Volts measure electrical *pressure,* or the amount of force behind each pulse. The higher the voltage, the farther a spark will fly from the pulsed wire to an animal's coat, and the greater the chance it will penetrate that coat. High voltage is necessary for stopping furry animals like coyotes and bears, hollow-haired animals like deer and goats, or thick-skinned animals like rhinos and elephants. On the other hand, since horses, pigs, and cattle have thin skin and relatively little hair, low voltage will keep them in.

Depending on the brand, American-made energizers range from less than 1,000 volts to over 10,000. Imported units range from about 3,000 to no more than 10,000. It takes 4,000 to 7,000 volts to impress most animals. A thick-coated animal, however, can bump against a 5,000-volt wire and never feel a thing, but if the same animal touches the wire with its nose, tongue, or ear, it will know it.

The more voltage a controller puts out, the more easily it throws sparks and the more readily it leaks energy through nearby weeds and faulty insulators. A high-voltage energizer therefore works best on relatively short fences. For long runs of fence, a lower-voltage controller will short out less readily.

Amperes or "amps" measure the *amount* of electricity flowing through a wire. Amps give a fence its shock value. If the amperage is too high, fuses will blow or circuit breakers will trip. It is the high amperage of makeshift homemade fence controllers that kills.

Watts measure the amount of *work* an electrical pulse does — its "horsepower," you might say. One horsepower, in fact, is equivalent to 745 watts. The number of watts in each pulse equals the amperage multiplied by the voltage.

You won't hear much about watts in a discussion on energizers, but you will hear the word "joule" bandied about, especially if you are looking at imported units. One joule is the amount of energy needed to produce one watt for one second. For comparative purposes, you need to know not only the joule rating but also how long each pulse lasts — the energizer's on-time.

For a given on-time, the higher the joule rating, the more watts the energizer puts out. If you compare two units with the same joule rating, but not the same on-time, the one with the longer on-time puts out fewer watts. If both units produce the same voltage, the one with the longer on-time puts out fewer amps and isn't as effective as the one with the shorter on-time.

As a general rule, a short-shock energizer requires 1.2 joules per mile (1.6 km) of line wire to control cattle and horses, 2.8 joules per mile for hogs, sheep, goats, and wildlife. Here, too, make sure the advertised rating is joules *delivered*, not joules *stored*. If your fence is too long for the energizer you want, divide it into separate sections and connect each to its own energizer.

Features

The number of different features an energizer can have is truly amazing. Some units have meters that tell you how much energy goes out with each pulse. Some have lights that, along with a steady "tock-tock-tock," tell you the unit is operating properly. Some have earth monitor lights that supposedly let you know the system is properly grounded (but you'd be wise to check the grounding system for yourself, as discussed in the next chapter).

Some units have two- or three-speed power control, letting you adjust how often a pulse goes out or how strong the pulse is. Use the higher setting while training livestock and during dry conditions; use the low setting during wet weather and after animals have learned to respect the fence. Reduce the setting, too, around horses and children, since both frighten rather easily when shocked.

The lower setting drains less power, an important feature for a battery-operated unit. On the other hand, if you have a serious predator or deer problem, stay with the highest setting. Some energizers have multiple terminals so you can run a longer fence on high power at the same time as you run a shorter barnyard fence from the lower-powered terminal.

The best energizers are guaranteed for at least two years and are covered against lightning strikes. Some units are highly prone to lightning damage and the power surges that have become a fact of modern life. Unless it is struck by lightning, a good energizer should last at least ten years, but will likely need repair every two to five years.

Battery units generally last longer than solid-state plug-ins because they aren't subject to power surges, the main source of deterioration. You can increase the life of any energizer by properly grounding your system, by using sound insulators and keeping them clean, by keeping your fence line cleared of weeds, and by installing a lightning diverter (as discussed in the next chapter).

Safety codes for plug-in fencers are established in Canada by the Canadian Standards Association (CSA) and in the U.S. by two organizations, The National Bureau of Standards and Underwriters'

Laboratories (U/L). Some states ban the sale of energizers that don't bear the "U/L Listed" label. All U/L-approved energizers have built-in lightning protection, and all will deactivate if something goes haywire so that too much current gets through for too long.

While all U/L-approved energizers are safe, an energizer that isn't approved isn't necessarily unsafe (although it *might* be). For one thing, Underwriters' Laboratories tests only plug-in energizers and only those submitted, along with payment, for approval. For another thing, any energizer that meets international safety standards (as specified on its label) is at least as safe, if not more so, than a unit with U/L approval.

TYPES OF ENERGIZER

Energizers are classified according to where they get their power. Plug-in, or "mains" units, work from 120-volt household (AC) current. There are four kinds of plug-ins: so-called "weed burners," continuous controllers, conventional controllers, and short-shock energizers. Battery-operated (DC) energizers come in two varieties: regular and solar. The storage battery of either may be dry cell, wet cell, or gel cell.

Some energizers can be run either from a wall outlet or a battery, as the need arises. Some plug-ins automatically switch to battery if the power kicks off, and will recharge the battery when power returns. A 12-volt DC converter with a built-in battery charger lets you manually convert nearly any battery-operated controller to a plug-in.

In general, plug-in units are the most reliable. Their operating cost is relatively low because of the low wattage they consume. Depending on the brand, energy use ranges from 20 to 250 kilowatt-hours (kwh) per year, averaging about 7 kwh per month. To determine how much a particular plug-in energizer will cost you to run, multiply its specified kwh rating by the kwh fee your local utility company charges.

Using a battery-operated energizer won't eliminate energy costs — you still have to recharge the battery periodically. Even if you recharge with solar energy, eventually the battery will wear out and need replacement.

Weed Burner

Weed burners, also called weed "choppers" or "clippers," send out pulses with a relatively long on-time — five to fifteen times longer than the 0.2-second maximum set by safety standards. Since the amount of current these energizers put out is reduced to keep them safe, the pulse travels only a short distance. The best unit can energize no more than about 5 miles (8 km) of line wire and issues such a weak pulse that it deters only the most timid animal.

The effectiveness of these controllers is reduced even further by weeds and wet weather. Besides all that, the circuit breaker, called a "pulser" or "chopper," requires frequent replacement. All you have to do is pull out the old one and plug in a new one, but you do need to keep spares on hand and check the controller often.

The chief advantage to a weed burner is that it puts out a big enough spark to singe back vegetation. In dry weather, though, that spark can start a grass or brush fire. When you hear about a fire started by someone's electric fence, you can be pretty sure a weed burner was involved. These controllers will also melt electroplastic wire (described in chapter 10).

Weed burners use three times more electricity than solid-state units, are not approved by any code, and are illegal in some states. Why does anyone buy them? Because they're cheap.

Continuous Controller

Another kind of energizer that is not U/L-approved (but *is* approved by the National Bureau of Standards) is the "continuous" controller, a modified form of weed burner. Unlike units that pulse current through line wires, this one keeps wires continuously charged. The feature can be dangerous because there is no off-time during which a small person or animal touching the fence can break away.

Continuous controllers therefore maintain very low current, between .003 and .005 amps. By comparison, a pulsed energizer with a 0.1-second on-time puts out between .025 and .040 amps.

The chief advantage to a continuous controller is that it lasts longer than a pulsating unit because it contains no delicate timing and circuit-breaking mechanisms. As a result of its very low power output, it is suitable mainly for clean, short fences no longer than 2 miles (3 km) used to protect gardens and confine pets. You will occasionally see continuous controllers packaged together with a few posts and a length of wire and sold as "pet kits."

Conventional Controller

The old-style high-impedance plug-in energizers are known as "conventional" or "standard" controllers. Most made today are solid-state, meaning they operate electronically and have no moving parts. Because their energy dissipates rather rapidly, they are suitable for energizing no more than about 20 miles (32 km) of wire. Despite their high voltage, often in excess of 10,000 volts, their relatively low amperage makes them ineffective for confining hard-to-control animals like sheep and goats and hard-to-deter predators like dogs.

Short-Shock Controller

A good plug-in, low-impedance energizer will cost you more than a conventional controller, but won't wear out as fast, will deter almost any animal or predator, and will effectively energize a much longer fence. It is also powerful enough to wither weeds. To avoid starting a fire in dry grass, select a model with an on-time of no more than 0.003 seconds.

Some units issue a steady weak pulse in conjunction with a random high-powered pulse designed to foil tricky critters who learn to deal with regular shocks. The problem with these is that predators don't always appear at the right time to get fully zapped.

Every solid-state energizer has a modular circuit board. Some are designed so you can easily replace them yourself. Others must be taken to a technician for repair, putting your fence out of commission for the duration unless you have a spare energizer. Circuit board damage is generally cause by insufficient grounding, discussed in chapter 8.

Battery-Operated Controller

Where electricity isn't available or power outages are frequent, a battery-operated energizer may be your best bet, although it can't handle much weed load. To avoid draining the battery too quickly,

these units pack much less wallop than a plug-in.

Least effective are throwaway dry cell models, including those operating on ordinary flashlight batteries. These inexpensive throwaway units are appropriate for portable fences, pet control, and seasonal gardens. They might be your best choice if you don't have a place to store things or you plan to move and don't need anything more to cart around. The most you can hope to energize is about 1 mile (1.6 km) of wire for no more than four months — not even that long, if weeds touch the fence.

A portable battery model may not be an effective energizer for most fences, but it does make a handy addition to any repair kit. When you troubleshoot your electric fence, disconnect its main energizer and carry along the portable unit to help you ferret out shorts as you work along the fence line. You can then easily disconnect the portable unit when you need to make a repair.

The bigger battery-powered energizer hooks up to a wet cell, 6-volt or 12-volt battery. The former — left over from the days when cars used smaller batteries — isn't nearly as effective as the latter. Higher-priced models come with an enclosed weatherproof compartment for the battery. The best units will energize up to 10 miles (16 km) of clean wire, but you will have to recharge the battery fairly often to maintain sufficient shock value.

Some manufacturers offer built-in gel pack batteries, which are smaller and easier to ship than a standard wet cell. They are considerably more expensive, though, and they have much less storage capacity.

Your best bet is a deep-cycle marine battery. In contrast to a standard shallow-cycle car battery, designed for continuous recharging, a deep-cycle battery is designed to be discharged fully before being recharged. A deep-cycle battery usually has a greater amp hour capacity than a car battery, too, so it will last longer between charges. You can find one at a boat shop or through Sears. The best have a built-in indicator to let you know how well the charge is holding.

How often the battery needs recharging depends on its quality and on how much power your energizer uses. You may need to recharge it more than once a week, or every six weeks. If you can lock your animals inside, recharge the battery overnight. Otherwise, recharge during the day when predators are less active.

Consider getting two batteries so you can have one recharged and ready when the other starts to fade. That way, you will never be caught short if a tree limb falls on your fence and drains the battery overnight. For convenience's sake, pick up your own portable battery charger, available from Sears or any automotive store. Use the trickle charge setting, since a fast charge may fry your deep-cycle battery.

Solar Recharging Controller

For a fence located in such a remote area that you can't check and recharge your battery often, use a solar panel to keep the battery charged. A solar panel can be hooked up with your battery system, or you can buy an energizer with a built-in solar panel and gel cell battery.

If you opt for a separate solar panel, be sure the power collected by the panel balances the power used by your energizer. If you live in the Northeast, the extreme Northwest, or any other area where sunshine is scarce, you may need two panels to keep your battery sufficiently charged, especially in the winter months.

The panel must be rated for a higher voltage than your battery, otherwise it can't fill the battery to capacity. On the other hand, if the panel is rated too much higher, it will keep trying to fill a battery that is already full, thereby cooking the battery. A 14-volt panel is about right for a 12-volt battery. Hooking up a voltage regulator is another way to keep the battery from getting overcharged, in addition to ensuring that it doesn't discharge through the panel at night.

If you choose a package deal complete with gel pack, the manufacturer has correctly matched panel input and energizer output. However, whether that match is right for you depends on the amount of sunshine that falls in your area — you don't want the battery to draw down faster than it can be recharged.

In any case, you will have to charge up the battery before connecting the energizer to your fence. Do so by leaving the unit in the sun, with the switch off, for the time specified in the instruction sheet — which could be as long as three weeks.

Although an electric battery charger works faster, it could damage the gel pack. If you don't plan to use the energizer for three months or more, store it away from sunlight.

Solar gel pack units tend to be at the weaker end of the battery energizer scale, offering little deterrent to hungry predators or livestock determined to break out. That is because the average gel cell stores only about 10 amp hours compared to 85 to 100 amp hours stored by a standard 12-volt wet cell battery, and the higher the energy output the quicker the battery draws down.

Even with a more powerful wet-cell-powered energizer, no two experts agree on whether it is best to use a standard shallow-cycle car battery or a deep-cycle marine battery. A deep-cycle battery isn't equipped for the constant discharging and recharging inherent in solar-generated power. Unless the battery is periodically discharged, it will eventually lose its memory and fail to hold any charge at all.

A shallow-cycle battery, on the other hand, *is* designed for constant discharging and recharging, but the long overnight discharge hastens its deterioration — much like drawing down your car battery by leaving the headlights on all night.

Solar panels depend on light, not heat, and will continue to recharge even in the indirect light of morning and evening, and on cloudy days. In fact, panels generally work better at lower temperatures. The more intense the light, though, the greater the recharge rate. With no light at all, some solar-powered battery energizers will work for up to three weeks, others as long as eight weeks. If you live in a rainy climate or other area that doesn't average at least five hours of light per day, forget solar.

ACCESSORIES

A few accessories are important to acquire at the time you get your energizer.

Fence Tester

A hand-held fence tester, or voltmeter, is indispensable for finding shorts in your fence. A good hand-held tester is rugged enough to withstand being carried in your pocket or tossed into your glove compartment. It is waterproof enough not to fog up in rainy weather, is easy to use, and gives a consistent read-out. Unfortunately, some testers are like today's bathroom scales — step on and off in rapid succession and you won't get the same reading twice in a row.

Digital voltmeters are calibrated in kilovolts (kv) and are rated for up to 9.9 kv. They're more accurate, easier to use, and easier to see than light-display testers. But use one only with a short-shock energizer, otherwise you will run the risk of burning out the tester. For a conventional controller, use a light-display voltmeter, also known as a neon light tester. Don't use any tester with a weed burner — you will definitely destroy it.

A tester has two terminals. One connects to a pulsed wire, the other to some part of the ground system. If the tester's ground terminal is a clip, attach it to a nearby grounding rod, a metal fence post, or a grounded line wire. If it's a spike, poke it into the soil. With both terminals in place, the tester will indicate how much voltage is getting through the pulsed wire. If you buy a cane-style tester, you can use it to ground pulsed wires while you make repairs.

ENERGIZER SELECTION GUIDE

| Type | Most Suited For | | | | | | Remarks |
	FENCE LENGTH	WEED LOAD	STOCK	PREDATOR CONTROL	GARDEN	HIGH-TENSILE	
Weed Burner	5 mi (8 km)	dry only	cattle horses hogs	no	yes	no	sparks fires; melts aluminum and electroplastic wire
Continuous Controller	1–2 mi (1.5–3 km)	none	horses pets	no	yes	no	unsafe—no off-time; low effectiveness
Conventional	15–25 mi (24–40 km)	moderate	cattle deer sheep	yes	yes	limited	melts electroplastic wire
Short Shock	50 mi (80 km)	high	cattle* deer goats sheep	yes	overkill	yes	controls difficult animals
Battery Powered (12 volt)	5 mi (8 km)	none	cattle goats horses sheep hogs	yes	yes	limited	requires battery charger
Battery Powered (6 volt or dry cell)	500 ft (150 m)	none	pets	no	yes	no	limited life
Solar Generated	10 mi (16 km)	none	cattle horses sheep	yes	yes	no	doesn't recharge in rain or shade

* including long-haired

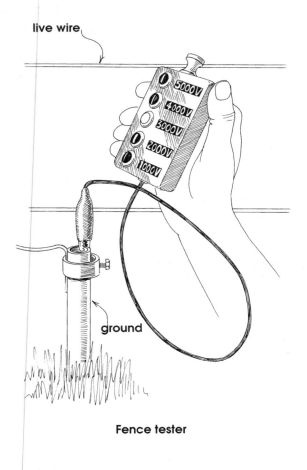

live wire

5000V
4000V
3000V
2000V
1000V

ground

Fence tester

Night Light

An option for worrywarts is a night light that connects to your fence and flashes every time a pulse travels through the wires; when you look out the window at night, this reassures you that your fence is working. The device, however, takes power from your fence.

Surge Suppressor

More than half of all plug-in energizer damage results from input power surges due to lightning,

erratic power supply, or large appliances on the same circuit as the energizer. You can avoid the latter by plugging electric motors or welders into a different circuit than your energizer uses. If your plug-in energizer doesn't have a built-in surge suppressor (most don't), buy one from any electrical supply center or Radio Shack. Plug the suppressor into the wall outlet and plug the energizer into the suppressor.

Protective Box

A battery-operated energizer and its battery, left out in the open, are tempting to someone inclined to take a potshot or load up the whole thing and haul it home. To prevent such mishaps, and to protect your controller from weather and curious animals, hide it inside an enclosed box.

Make the box of wood, with a sloped roof to shed water; use an oversized mailbox; or buy a box made of galvanized steel designed for the purpose. The nice thing about the latter is that it becomes energized when you lock the door with an insulated key. Anyone who comes along with larceny at heart is in for a big surprise. Take care, though, as some metal boxes draw a huge amount of current from the fence. Make sure that you can return the thing if you're not happy with it.

Warning Signs

In some areas the law requires you to post warning signs at 200- to 300-foot (60 to 90 m) intervals along any electric fence. Bright yellow signs printed with the words "Electric Fence" are sold by farm stores and suppliers of fencing materials. Your county extension agent should be able to tell you exactly where and how signs must be posted to comply with local laws. Even if the law doesn't require it, posting signs puts potential trespassers on notice.

CHAPTER 8

INSTALLING AN ENERGIZER

▼▲▽▲▽▲▽▲▽▲▽▲▽▲▽▲▽▲▽▲▽▲

Set up your energizer before stringing the first wire of your electric fence. Then, if you don't get the fence finished in one day, when you quit for the night you can electrify the portion that is done to minimize the chance that wandering wildlife will get tangled in the newly strung wires.

Install and operate your energizer according to the manufacturer's instructions as well as local and state regulations and the National Electrical Code. Information regarding legal codes is available from your area's building inspector, an electrical contractor, or maybe your local library.

INSTALLATION

Installing an energizer involves securely mounting the unit on a wall or post, attaching the hot terminal (sometimes labeled "fence" or "positive") to fence wires, and attaching the ground terminal (sometimes labeled "earth" or "negative") to a grounding rod.

Lead-out wires for this purpose are sometimes provided with the energizer, but they are often too short or too thin and should be replaced. To avoid confusion, follow standard convention by using red or black insulated wire for the hot lead-out and green or white for the ground.

Use only 12½-gauge insulated galvanized steel wire, available from electric fence suppliers. Anything thinner will reduce your fence's conductivity right at the power source. Never use insulated household copper wire. It is designed for indoor use and is rated for only 600 volts — much less than the voltage of most energizers.

If your energizer isn't located at the beginning of your fence, run the lead-outs underground. Placing them on the ground means you will constantly trip over them. Stringing them overhead means they will get snagged in equipment or vehicles passing below and will attract lightning strikes.

Protect underground wires with insulated tubing, available from electric fence suppliers. If heavy traffic will pass over them, slip the wires inside a ½-inch (13 mm) pipe. Since trapped moisture can create problems, orient both ends of a flexible pipe downward by folding each back on itself and taping it. If the pipe is rigid, seal the ends around the emerging wires with silicon.

Where the distance from the energizer to the start of the fence is more than 65 feet (20 m), connect two lead-out wires side by side to minimize voltage drop. Where the distance is more than 325 feet (100 m), use three wires.

Connect lead-out wires to the energizer with standard circular compression fittings, available from Radio Shack or any electrical supply outlet. If the wires supplied with the energizer are fitted with alligator clips, replace them with compression fittings to improve contact.

To connect the lead-out from the energized terminal to the fence, use a compression sleeve, a twist connector, or any galvanized solderless connector such as a wire tap or screw-down clamp. Any of these will give you a firmer connection than simply wrapping the lead-out around the line wire. Screw-down clamps are particularly handy because they can easily be disconnected, should the need arise.

Connect the lead-out from the earth terminal to a grounding rod by fastening it tightly with a proper ground clamp that bites into the rod. There are two kinds of clamp to choose from, depending on whether your wire and ground rods are of similar or dissimilar metals. When you buy the clamps, be sure to get the right kind to avoid early corrosion that will interfere with your grounding system.

Energizer Position

The energizer's label will tell you whether or not the unit can be left outside. Plug-ins must be kept inside; battery models are usually designed for outdoors use. Indoors or outdoors, mount your energizer where curious animals can't accidentally disconnect it, and out of sight of vandals and thieves.

Some outdoor models come with a mounting stand. Others can be attached to a post or hidden inside a locked box. An energizer requiring indoor protection must be mounted in a clean, dry place, securely attached to the wall, never left on the floor or on a shelf. If lead-out wires must pass through a wall, thread them through plastic pipe or insulated tubing (described in chapter 9). Never let the wires touch the building or each other.

You don't necessarily have to place the energizer where you can connect both ends of the fence to it in a closed loop. Doing so, however, ensures that no part of the fence will be disabled if a short circuit occurs anywhere along the line. The chief disadvantage to a closed loop is that shorts are harder to detect.

With your fence connected to the energizer only at one end, you can use your voltmeter to compare the voltage at the start to the voltage at the end. A drop tells you to look for shorts. Actually, the end of the fence needn't come back to the energizer at all. You might run the fence in a straight line if you are using it, for example, to divide a field enclosed by a perimeter fence.

If your fence is too long for your energizer, line wire resistance will cause the voltage to drop gradually along the way. How long your fence can be before voltage drop becomes a problem will be determined by a number of variables, including the type of wire you use, the number of strands, and the strength of your energizer — the best you can do is take an educated guess based on your particular circumstances and then adjust and fine-tune as necessary. Overcome resistance by dividing the

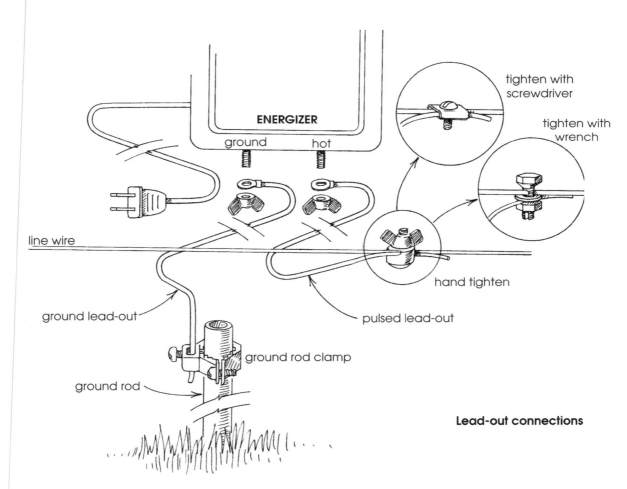

Lead-out connections

fence at midpoint to reduce the distance the current has to flow. If that doesn't do it, separate the fence into two sections and use two energizers. Never connect more than one energizer to the same section of fence or you could increase the current and the on-time to dangerous levels.

Grounding

The success of an electric fence depends on proper grounding, also called "earthing." Grounding literally means providing current a ready path through the soil. Surveys here and abroad show that at least 80 percent of all electric fences have inadequate grounding systems.

Poor grounding can cause voltage leaks to get back to the energizer, blowing fuses or seriously damaging the controller. If you operate a dairy and use milking machines, poor grounding can cause your equipment to draw current from your energizer — much to the discomfort of your goats or cows.

For an electric fence to work, current must flow from the energizer through the line wires and back to the energizer through the soil. Under normal circumstances, electrically charged wires are separated, or insulated, from the earth and the circuit is open. A person or animal standing on the ground and coming into contact with a pulsed wire closes the circuit — and feels the consequences.

A bird landing on a hot wire doesn't feel a thing because it doesn't complete the circuit. If, while sitting on the wire, the bird pecks at a metal fence post, it's bye, bye birdie. One of the hazards of electric fencing is that lizards, praying mantises,

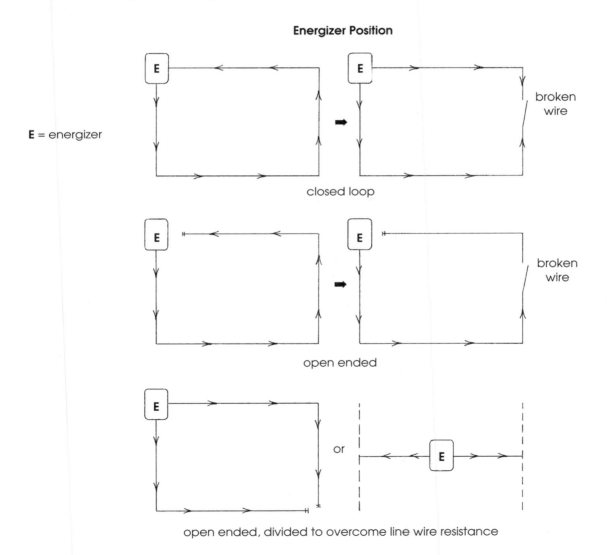

Energizer Position

E = energizer

closed loop

open ended

broken wire

broken wire

open ended, divided to overcome line wire resistance

and other small creatures get fried when they crawl up a metal post and reach for a hot wire, thus closing the circuit. The same jolt that is lethal for creatures of small mass, though, is merely unpleasant to a larger animal.

In order for that animal to complete the circuit, the energizer must have good contact with the soil. Contact is provided by means of a ⅝-inch-diameter (16 mm) grounding or earthing rod made of carbon steel, galvanized, copper, or copper clad to resist corrosion. Rods are sold at electrical and building supply outlets. In some areas, building codes dictate their length and diameter.

Some fencers use ¾-inch (19 mm) galvanized steel pipe, which doesn't last as long as a proper earthing rod and is more difficult to connect firmly. In any case, never recycle a rod or pipe that is rusted or painted — the coating will reduce its effectiveness as a ground.

Since moist soil is more conductive than dry soil, the grounding rod must be driven deep enough to make good contact with damp earth. In many areas, that means 8 feet (240 cm). Depending on your soil moisture, you may get away with a shorter rod — provided it is adequate for your driest season, not your wettest. A rod that is too short won't let the energizer deliver its best spark.

The ideal spot for an earthing rod is a shady place that is moist year-round, such as the north corner of a barn where rain runs off the roof. If the best place for the rod is some distance from your energizer, connect the unit's insulated ground lead-out to an underground length of bare number 2 copper ground wire. Use a copper split bolt connector (available from electrical supply centers) to attach the insulated wire to the bare wire at ground level. The underground copper wire will not only provide a means of connecting the ground rod, but will also improve the entire grounding system through increased soil contact.

In a dry area, improve soil moisture by forming a depression to collect runoff. In this era of perennial droughts, you may have to soak the rod periodically to keep your fence working properly. To minimize evaporation, spread several inches of sand around the rod.

Even though moist soil is ideal for grounding, don't place rods in a running stream or along a stream bank — you could end up with a dead fence and a live stream. Keep rods away from underground metal pipes and metal structural supports in or on the soil and at least 20 feet (6 m) from the earthing rods of other systems including power and telephone poles. Check local codes — you may be required to keep rods as much as 60 feet (18 m) from existing utility ground rods.

Don't be tempted to use utility line earthing rods or your plumbing system to ground your fence. Besides violating code, you could cause interference with telephone, television, or radio reception, or suffer shocking consequences when you take a shower.

You will need at least two grounding rods for a conventional controller, as many as seven for a high-energy controller. If you are installing an imported energizer, the rule of thumb is to allow at least 3 feet (90 cm) of rod per joule of energizer output. In practical terms, that doesn't mean you have to pound in a 15-foot (450 cm) rod for a 5-joule energizer. If you're using 8-foot (240 cm) rods, round *up* to the nearest multiple of eight and use two 8-foot (240 cm) rods. If 5-foot (150 cm) rods are sufficient for your soil conditions, you will need three.

Place the first rod as near as possible to the energizer, keeping it *outside* any building or shed to minimize the chance of starting a fire or electrocuting livestock should lightning strike. Drive it in with a post driver. As the exposed portion gets shorter, finish it off with a sledgehammer. Leave about 6 inches (15 cm) above ground.

Space additional rods at least 10 feet (300 cm) apart and connect one to the other in series with copper ground wires, making firm connections with ground clamps. Since the rods stick out of the soil and have wires running between them, keep them close to the fence line to prevent damage by traffic or mowers, and put them on the side opposite livestock to prevent injury. For a garden fence, place the rods where you're least likely to trip over them.

To check your earthing system, drive a test rod into moist soil at least 500 feet (150 m) down the fence line from your energizer. With the energizer turned off, connect one or more hot wires to the test rod. When you turn the energizer back on, you should have what is known as a "dead short," meaning all the energy rushes down the wires through the test rod into the soil and tries to get back into the energizer via its ground rods.

To make sure you have enough grounding rods for that to happen, poke a large nail into the soil near the last rod in series. Attach the positive terminal of your voltmeter to the nail and the negative terminal to the last rod. If you get a reading, your grounding system is inadequate. Add more rods until the reading drops to zero. Don't forget to disconnect the test rod when you're done.

Retest your system at least once a year, preferably during your driest season. Before adding more rods, make sure all existing ground connections are free of corrosion.

HOT WIRES

Just as important as establishing an adequate earthing system is making sure the energized line wires have no contact with the ground. A direct connection would, of course, result in a dead short. Lesser contacts, called "shorts" or "leaks," drain energy from your fence, making a plug-in system less effective and drawing a battery system down more rapidly. A leaky electric fence is like a leaky garden hose: neither delivers pressure where it's needed.

Energy leaks have any number of causes: grass or weeds growing against hot wires or wind-blown twigs or branches lying across them; cracked insulators; bugs, leaves, or dust lodged in the insulators; salty sea air; or a broken wire flapping in the breeze.

Most of these situations cause "arcing." Pulses of current, instead of flowing smoothly from one place to another, jump across a narrow gap by means of a spark. When arcing occurs, you can usually hear the "snap" of the jumping spark, which can help you find and correct the source of leakage. Whenever you check your electric fence, therefore, listen as well as look for problems. If the source of the snapping sounds eludes you, go back at night and you will actually see sparks fly.

Earth Return

When a fence's circuit is completed by an animal standing on the ground, the system is called a "ground return" or "earth return" system. This system works well for relatively short runs of fence in areas of even rainfall where the soil is conductive and predators such as the wily coyote are not a problem.

Under certain circumstances, though, the earth-return system doesn't work well at all. Extremely dry soils, for example, are not very conductive. In winter, the earth may be insulated with a layer of packed snow. Sandy, rocky, or frozen soil is self-insulating.

The problem is compounded if you're dealing with hoofed animals, especially small ones. Hooves have more insulation value than soft pads. Animals with tiny hooves, like lambs and kids, are not only insulated from the ground but have very little contact to start with. In addition, young stock and some predators get through a fence by leaping between the wires. They aren't shocked because their feet are off the ground.

In any of these situations, an earth-return system isn't nearly as effective as a wire-return system.

Wire Return

The wire-return system, also called a "ground wire circuit," does not depend on weather and soil conditions. Animals that have built-in insulation such as hooves or thick fur and those performing tricky maneuvers such as jumping through a fence instead of crawling under or climbing over are de-

EARTH RETURN VS. WIRE RETURN

Earth Return	Wire Return
Animals touch only one wire	Animals must touch two wires
Requires good soil conductivity	Independent of soil conductivity
Fewer wires needed	More wires needed
Subject to fewer shorts	Subject to more shorts
Easier to maintain	More difficult to maintain
Works best over short distances	Works even over long distances

terred. Unlike the earth-return system, it works even for long fences.

The standard wire-return system involves connecting every other wire to ground. When an animal touches two adjacent wires, one hot and one grounded, it completes the circuit and feels a jolt. Even if the animal hits the fence with all fours off the ground, it must touch only two wires to close the circuit.

You will occasionally run across ads for so-called "bi-polar" energizers, mysteriously claiming to send out two sets of oppositely charged energy pulses. Such an energizer is nothing more than a built-in wire-return system — you hook up all the hot wires to the positive terminal and all the grounded wires to the negative terminal.

Exactly which wires should be grounded is a matter of debate among proponents of electric fencing. Some start with the bottom wire to reduce energy leakage from encroaching weeds. Others start with the second wire up as a better deterrent to young stock and small dogs. With the bottom wire hot, an animal pushing under the fence has two ways to complete the circuit — by touching the earth and the bottom wire or by touching the two lowest wires.

Also under debate is whether to earth the top wire. Some feel that doing so reduces the risk of lightning damage to the energizer. Since lightning strikes the highest point, if that point is a grounded top wire, the strike will be diverted directly to earth with little damage done.

Fencers who live anywhere but on the open plains argue that grounding the top wire does little to reduce the chance of lightning damage. They point out that lightning, rather than striking a fence directly, is more likely to strike a nearby tree and arc to the fence, thus hitting more than just the top wire.

Besides, if the top wire of an electric fence is nose-high to the animals being controlled, that wire should definitely be pulsed. If a grounded wire is desired for lightning control, an additional wire should be added above the top hot wire. Alternatively, the top wire could be attached so it can be easily switched from hot to ground during storms.

In developing a wire-return system that works for you, keep these rules in mind: energize the line wire nearest the animals' nose height; keep the hot and the ground systems separate from each other; string hot wires and grounded wires at least 4 inches (10 cm) apart so they can't inadvertently touch.

For a wire-return fence over 1,000 feet (300 m) long, you will need additional grounding rods, 6 feet (180 cm) long. If your fence is very long and your soil is reasonably moist, space the rods every 3,000 to 5,000 feet (1 to 1.5 km). If your fence is short and/or conditions are dry, space rods every 1,000 to 1,500 feet (300 to 450 m).

Earth-return system

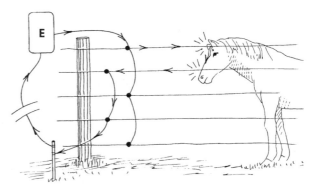

Wire-return system

Don't worry as much about spacing the rods evenly as putting them in low spots where the soil stays moist or the grass is greenest. Place each rod near a post, where it will be out of the way when you mow. Connect each rod to all the earthed wires, using the same green insulated wire you used for lead-outs.

These extra connections mean a wire-return system is subject to more shorts and is more difficult to troubleshoot than an earth-return system. It therefore makes good sense to maintain the latter as long as it works and switch to the former only as required by seasonally changing conditions.

Neutral Wires

Another way to accommodate seasonal variations is to make one or more bottom wires capable of being neutral — neither energized nor earthed. The advantage with neutral wires is that they can't mistakenly be connected to both the energizer and the ground, creating a dead short. The disadvantage is that, except when they *are* energized, they don't contribute to the effectiveness of an electric fence except as a physical barrier of sorts.

Seasonal deep snow and heavy weed growth are the two main reasons to consider stringing neutral bottom wires. Weeds and wet snow both draw current from the fence. Disconnecting the weed- or snow-covered wires improves the effectiveness of the wires above.

During a heavy snowfall, disconnect the lower wires, one at a time, as drifts deepen. In early spring, when weed growth is slow but lambs or kids are in season and predators are feeding their own young, reconnect the bottom wires. Later, when weeds grow faster than you can mow, disconnect the bottom wires to encourage livestock to graze closer to the fence. Since stock constantly test a fence, they will soon discover that close grazing is safe.

Rather than manually connecting and disconnecting wires, connect them by means of an energy limiter, more commonly used for floodgates as described in chapter 13. When an energy limiter senses serious leakage, it switches off the current to keep the leaks from draining the rest of the fence.

Series vs. Parallel

The trickiest part of electric fencing is getting all the line wires properly connected. When you decide which wires will be hot and which will not, connect together all the hot wires, then connect together all the grounded wires.

Line wires can be connected either in series or in parallel. Since each section of fence between anchor posts is constructed separately, it is easy enough to energize the sections in series. Connect the wires of each section separately, then join two sections by running a "feeder" or "jumper" wire from one hot wire of one section to one hot wire of the next section. Then run a feeder wire from one grounded wire of one section to one grounded wire of the next section.

Connecting in series has the advantage that by disconnecting the jumper wire you can isolate each section for troubleshooting. The disadvantage is that you can't easily switch individual wires from pulsed to earthed and back again.

Wires connected in parallel, on the other hand, can be easily switched back and forth from energized to earthed as the need arises. Parallel connections simplify the spotting of shorts and eliminate the chance that one broken wire will disable the whole fence.

Connect each line wire of one section only to the corresponding line wire of the next section. Each hot wire is attached directly to the energizer, each

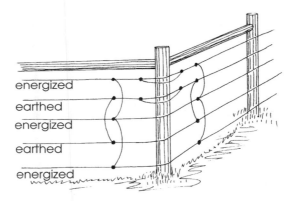

energized
earthed
energized
earthed
energized

Connected in series

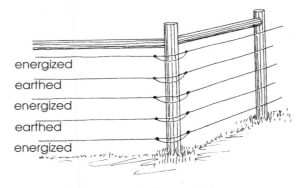

energized
earthed
energized
earthed
energized

Connected in parallel

earthed wire is attached directly to ground, and all wires are kept separate from the others. To trace a short, disconnect one wire at a time from the energizer and take a voltage reading on the rest. When the voltage goes up, you will know the disconnected wire is the troublemaker.

If your feeder wires are the same 12½-gauge insulated wire you used to connect the energizer lead-out, and you maintain the convention that red stands for hot and green stands for ground, shorts will be fewer and easier to trace. Remove insulation from the jumper wires at each point of contact, and scrape the line wire clean and bright.

Some fencers prefer to use regular galvanized fence wire protected with insulated tubing (available from electric fence suppliers). In any case, insulation is essential to ensure that a hot feeder doesn't touch any neutral or earthed wires, and vice versa. If you are constructing a tension fence, connect feeder wires only after full tension has been applied to avoid overstretching them or tearing them loose.

To make connections, you can use wire taps or screw-down clamps, but compression sleeves are cheaper as long as you remember to slip them onto each line wire when you erect the fence. Alternatively, use curl-on connectors, which double nicely as cut-out switches.

CUT-OUT SWITCHES

The longer your fence is, the more important it becomes to include cut-out switches that let you turn off the current at certain points along the way. Otherwise, when you spot a short you will have to walk to the energizer to turn it off, go back to the fence and make your repair, walk back to the energizer and turn it on, and return to the fence to continue your check.

You could, of course, carry rubber gloves to insulate your hands while you make repairs, but you still run the risk of klutzing into a live wire with some other part of your body. A voltmeter that lets you short out a wire while making repairs is a handy option, since you have to carry a meter anyway while checking your fence. Another possibility is to disconnect the main controller while you're trouble-shooting and energize the fence with a portable battery-operated unit.

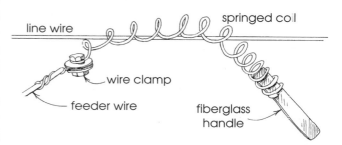

Curl-on connector

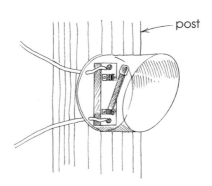

High-voltage knife switch

If you are into high tech, buy a remote control cut-out switch to attach to your energizer. It has a pocket unit you touch to a hot wire to turn off the power. Touch again, the power goes back on. The pocket unit has a built-in neon light voltmeter.

Feeder wires attached with curl-on connectors can easily be disconnected and reconnected. A curl-on consists of a coil of 12-gauge wire with a wire clamp at one end and a fiberglass handle at the other. Attach the clamp to the feeder wire and hang the coil over the line wire so that it makes contact in several places. The insulated fiberglass handle lets you remove the coil without getting zapped.

The standard way to turn off juice to a section of fence is with a high-voltage switch, sometimes called an "isolator." Be sure to specify high-voltage, since a low-voltage switch will cause arcing. Also, get switches designed for outdoor use or provide some rudimentary form of weather protection to keep them functional.

To turn the juice on and off, all you need is a "single pole" switch. If, in addition, you want to change selected wires back and forth from hot to ground, you will need a "double pole" switch. Both kinds are available from electric fence suppliers and electrical supply centers.

You can place a switch at any post on any nontensioned feeder wire. Simply cut the feeder, strip back a bit of insulation from the two cut ends, attach the ends to the switch's two terminal screws, and screw them down tightly.

The most convenient places for cut-out switches are gates and corners. Besides using them to isolate sections of fence, you can install switches on your lowest live wires so you can switch off the current when snow piles up or grass grows tall.

LIGHTNING DIVERTERS

One of the greatest threats to an energizer is damage resulting from lightning. An energizer struck by lightning can blow up like a hand grenade—with similar consequences. Even if it doesn't explode, it may be damaged beyond repair.

When electrified line wires are well isolated from the earth, current generated by lightning can travel a great distance to the energizer. Lightning diverters, spaced periodically along the fence, give the high-powered current an alternative place to go.

Diverters are sometimes mistakenly called lightning "arresters," but they don't really stop or arrest anything. Rather, they channel or divert the lightning's current to the earth, where it truly is arrested.

A lightning diverter consists of a 4- or 5-inch (10 or 13 cm) porcelain base with two little peaks or towers, about 2 inches (5 cm) apart, each ending in a galvanized bolt. One tower is connected to hot wires, the other to ground.

The diverter brings the energized system and earthed system as close together as possible without making contact. Since lightning has extremely high voltage, it can easily jump across the gap from the hot side to the earthed side, and thus be harmlessly diverted to the ground. If a diverter arcs, either your energizer is too powerful for your fence or the diverter is faulty and should be replaced.

To minimize the chance that lightning will strike between the diverter and the energizer, place the first diverter as close as possible to the energizer. Space additional diverters at least every mile (1.6 km) along the fence. In areas where severe lightning is frequent, place diverters every half mile (.8 km). If your fence measures ½ mile or less, put a

diverter at each corner.

Screw or bolt the diverter to a fence post, oriented vertically so one tower is above the other. Attach fence wires with the same insulated wire used for lead-outs. If lightning does strike, the insulation will melt and these wires will have to be replaced. Meanwhile, though, insulation keeps the fence from inadvertently grounding out.

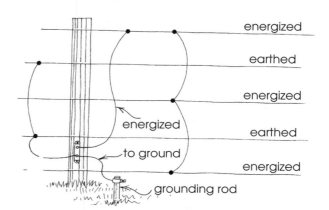

Lightning diverter

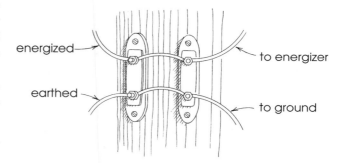

Two diverter fail-safe

Attach a length of 12½-gauge red insulated wire to the top terminal and connect it to all the hot wires. If your hot wires are in series, you will need a diverter for each wire. Some people connect only the top hot wire, but if lightning arcs from a nearby tree it can hit other wires as well. With a length of green insulated wire, connect the other terminal to all the grounded wires, if any, and to a grounding rod.

If you have installed grounding rods for a wire-return system, use them for the lightning diverters. Some fencing experts warn against this practice because too many fencers cut corners in setting up their grounding systems. As a result, the first powerful surge that comes along travels down the ground wires and blows up the energizer. It is far cheaper to buy separate grounding rods for the lightning diverters than to buy a new energizer. If you opt for extra rods, install one more for each diverter than you used for the energizer.

Lightning diverters aren't 100 percent effective. All the precautions in the world can only reduce the danger of lightning damage, never eliminate it. Since most electric fences offer little or nothing by way of a physical barrier, keep a spare energizer on hand if you live in an area of frequent, intense storms.

Lightning diverters can't protect your energizer from a direct strike or from lightning coming through the power lines. You can't do much about a direct strike, but you can protect a plug-in energizer by disconnecting it in severe storms — not a bad idea, since 98 percent of all lightning damage occurs through power surges.

It is not safe to be near an energizer with a storm overhead, so disconnect at the first sign of a pending storm. As a safety precaution, have the energizer plugged into a switched outlet. Then you can turn off the unit without touching it. Just remember to flip the switch back on after the storm passes.

Lightning Brake

Lightning generates such powerful force that some of its energy could continue down a line wire after only a portion is diverted to the ground. As a fail-safe measure against a surge that may exceed the capacity of one diverter, place two diverters, side by side, in the position closest to the energizer.

For a diverter to do its job, lightning must have a good incentive to jump the gap between the towers. Such incentive is provided when a diverter's ground system is better than the energizer's ground system. The diverter closest to the energizer should have more grounding rods than the energizer has, and the two sets of rods should be at least 20 feet (6 m) apart.

Additional incentive is offered by a lightning brake, also called a "choke," "resistor coil," or "induction coil." This device, placed between the hot wire lead-out and the first diverter, acts as a magnet-

ic brake, increasing resistance and encouraging the high-voltage current to jump the towers in seeking the easiest path to ground. Induction coils are available from electric fence suppliers and from Radio Shack. You could also make your own.

To make a simple brake, wind 20 feet (6 m) of 12½-gauge insulated wire around a 10- or 12-inch-diameter (25 or 30 cm) bucket until you have six or seven loops (depending on the size of the bucket). Tape the coil together with electrical tape and slip it off the bucket.

For a brake that doesn't rely on tape to hold it together, you will need two lengths of plastic pipe 14 inches (35 cm) long with holes drilled through at 2-inch (5 cm) intervals. Temporarily secure the pipes parallel and 12 inches (30 cm) apart. Thread 12½-gauge fence wire between the holes of one pipe and the other, as illustrated, making five complete loops. At the last hole on each end, wrap the wire once around the pipe to hold it in place.

Securely connect one end of a lightning brake to the energizer's red lead-out wire and the other end to the fence's hot wire system, making certain there is no other path to the energizer except through the

Lightning brakes

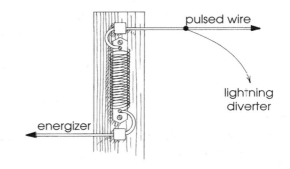

pulsed wire

lightning diverter

energizer

Store bought

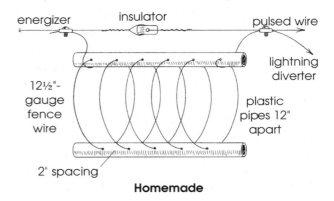

energizer

insulator

pulsed wire

lightning diverter

12½"-gauge fence wire

plastic pipes 12" apart

2" spacing

Homemade

coil. Whether you need a brake or any other precaution against lightning depends entirely on the frequency and severity of lightning storms in your area.

TROUBLESHOOTING

When your electric fence is all hooked up, check it with a voltmeter before introducing livestock. After spending a tiring day building the fence, it is easy to overlook one connection or mistakenly attach a hot wire to ground, causing a dead short. To be fully effective, the fence should carry at least 4,000 volts or 4 kilovolts.

Make a note of the voltage reading so you will have a point of comparison for troubleshooting later

on. Recheck the voltage at least weekly, preferably daily. Make it part of your regular routine. Livestock and predators constantly test your fence, and so should you. Check, too, after every lightning storm or deep snowfall, and any time you make a change or repair.

Don't worry if the voltage drops during or immediately after a shower. Rain increases conductivity and causes an instant decline in voltage. However, if the voltage stays low after things dry out, start looking for the source of trouble.

If you use an earth-return system, take readings at several places at different times of year so you will know what to expect and can detect a significant drop. To check an earth-return system, connect the fence tester to a hot wire and a ground rod, metal

USING A FENCE TESTER

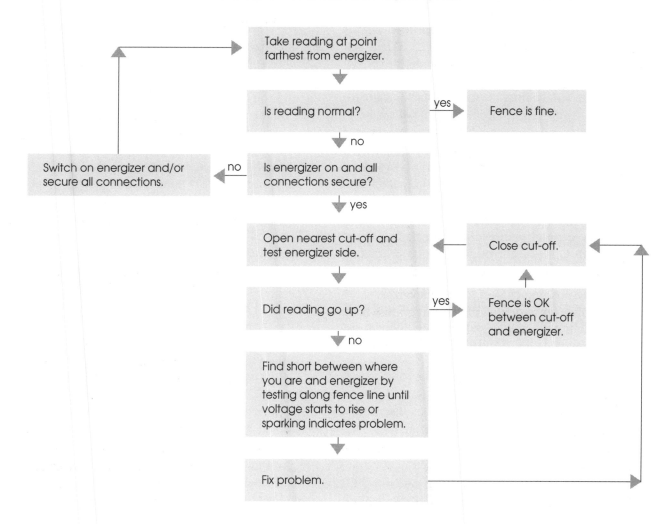

fence post, or piece of stiff wire poked 6 inches (15 cm) into the soil. To check a wire-return system, connect the tester to a hot wire and an earthed wire, then connect it to a hot wire and a ground rod, metal post, or length of stiff, earthed wire. The first reading should be equal to or greater than the second.

To ensure current gets through the full length of your fence, start testing at the end farthest from the energizer. Check each hot wire, paying particular attention to those at heights your animals are most likely to touch. If the voltage is low, continue along the fence, testing at intervals and looking for a short. If the voltage suddenly goes up, you have passed the short without spotting it.

What to Look For

As time goes by, you will develop intuition about what to watch for. Besides physical destruction from the environment (a fallen tree or a mud slide, for example), the two main things to look for are short circuits and open circuits. An open circuit means a hot wire got loose or broke. When the circuit is open, no current passes through.

A short circuit means one or more live wires have come into contact with the earth, causing current leakage. Short circuits are of two sorts. A high-leakage short, or dead short, happens when a hot wire touches the earth, a grounded wire, or a metal fence post. A low-leakage short occurs when a hot wire touches something less conductive, like weeds or a wooden fence post.

Other things to watch for are line wire resistance and faulty ground connections. Resistance may result from loose or corroded connections or line wire that is too long, too thin, or too rusty. Corrosion and rust occur over time, but resistance as a result of long, thin wires will show up the first time you take a reading. Loose ground connections are difficult to detect unless you are willing to spring for very expensive equipment.

Any time you have to make a repair, disconnect the part of the fence you are working on. If you do so by turning off the energizer, either disconnect the feeder wires or develop some system for notifying family members that it is *not* being helpful to turn the energizer back on. Make sure everyone, especially children, knows how to turn off the energizer in an emergency. Powering a plug-in through a switched wall outlet lets anyone turn the thing off without studying control switches.

There is always a chance the energizer itself will malfunction. Moisture, dirt, or insects getting into the electric circuit can ruin any controller. If it's a plug-in, verify that the power is on and the unit is plugged in. Check the fuse and replace it if necessary — if it blows again right away, the energizer needs repair. If it blows within a few hours, your ground system is at fault. If the power is on and the fuse hasn't blown, connect your voltmeter to the hot and ground terminals. A low reading means the energizer needs repair.

If you have a battery unit, the battery may be dead or its terminals corroded. The latter is fairly common, as it takes little corrosion to stop the flow of current. Prevent corrosion by keeping terminals thinly coated with petroleum jelly. If corrosion does occur, wire-brush the terminals and recoat them.

Interference

An energizer can interfere with radio, television, CB, or telephone reception by causing short, steady bursts of static. How bad the interference gets depends on the energizer's brand, its distance from your house, whether it is properly grounded, and how weak your radio or TV signal is. You may get no interference with network news, but switch to PBS and the blips could drive you wild.

To minimize interference, avoid running fence wires parallel to or within 24 feet (7 m) of overhead or underground telephone lines, incoming cables, and radio, TV, or CB antennas. If your energizer is a plug-in and doesn't have a built-in surge suppressor, adding one between the plug and the wall outlet *may* reduce the problem.

Radio, TV, or CB interference is usually caused by a faulty connection in the fence, most often in the ground system. Make sure all connections are secure and check for cracked or debris-filled insulators. Interference of this nature usually gets worse during wet weather.

If your energizer is a plug-in, interference could be caused by the unit itself or by faulty electrical wiring. To determine if either is the case, disconnect the energizer from the fence and from ground. Turn on a transistor radio and tune it between stations at the highest frequency. If you don't hear any clicks, interference is the result of a poor connection in the fence or its ground system.

If you hear clicks on the radio, the problem is either in the energizer or the electrical wiring. To find out which, have someone at the house listen to the radio or TV getting interference. With the energizer disconnected from the fence and from ground, turn it on and, after it clicks a prearranged number of times, pull the plug.

If the person at the house hears the same number of clicks that you do, the problem is in the power supply. Try plugging the energizer into another socket. If that doesn't work, check the radio's or TV's plug and grounding system. If the person at the house hears one more click than you do, your energizer needs repair.

TROUBLESHOOTING YOUR ELECTRIC FENCE

If this occurs:	Ask yourself:
No energizer output	Is the power on? Has a fuse blown? Are battery terminals clean? Does battery need recharging? Does battery need replacement?
Low energizer output	Is switch defective? Are switch connections loose? Are battery terminals clean? Does battery need recharging? Does battery need replacement? Does energizer require repair?
No voltage at fence	Is lead-out firmly connected? Is lead-out wire corroded? Are ground rods connected? Are jumper wires connected? Are ground wires connected?
Low voltage along fence	Is energizer powerful enough? Are battery terminals clean? Does battery need recharging? Does battery need replacing? Is ground system connected? Is ground system adequate? Is ground-return wire needed? Are insulators dirty or broken? Are too many weeds encroaching? Has anything lodged in wires?
No voltage at some point	Is a jumper wire loose? Is a hot wire broken? Is a hot wire grounded?

CHAPTER 9 ▽▲▽▲▽▲▽▲▽▲▽

ELECTRIFIED FENCES

▽▲▽▲▽▲▽▲▽▲▽▲▽▲▽▲▽▲▽▲▽▲▽

By providing a psychological barrier rather than a physical one, an electric fence operates by fear rather than by force. It will control even the most difficult livestock or predators, including bulls, bears, coyotes, dogs, deer, feral pigs, and goats.

Despite these advantages, electric fences are not suitable for all situations. You would never use one for a closely confined area where animals crowd each other into the fence. You certainly wouldn't string one above or close to a watering trough or ye olde swimming hole. You should never run one under a power line — if the line came down in a storm, fence wires could transmit dangerous levels of current. Finally, in some areas the law prohibits electric fences adjacent to roadways or public lands. Check your local ordinances.

WIRE

To be effective, an electric fence relies not just on a powerful energizer, but also on the conductivity of its line wires. Conductivity is a measure of how readily current flows. Resistance is the opposite of conductivity. Highly resistant materials are used in making insulators to isolate electrified line wires, or conductors, from the earth.

The longer the fence, the more important it is to use good conductors and good insulators so current can travel the full length of the fence with minimal leakage. Today's electric fences work better than yesterday's not only because energizers

are better, but also because we have better conductors and better insulators.

Wire for an electric fence is selected on the same basis as wire for a nonelectric fence — the material it's made of, how it's coated, its thickness, and its tensile strength. Here, though, an important consideration is the effect of these four properties on the wire's conductivity.

Tensile strength doesn't directly affect conductivity, but it does influence the effectiveness of an electric fence. For one thing, high-tensile wire is less likely to break, and a broken wire means a dead fence. For another thing, high-tensile wire can be strung tauter so is less likely to sag and touch something that might cause a short.

Because it can be strung so tightly, high-tensile wire also offers a physical deterrent to livestock tempted to slip out or predators tempted to slip in. Although it requires stronger anchor assemblies than soft wire, high-tensile wire lets you get by with fewer line posts—making the whole fence cheaper.

Conductivity

The other three properties — base material, coating, and thickness — directly affect conductivity. Since electricity moves down the outside surface of a conductor, the thicker the wire, the more surface it has and the higher its conductivity. Wire of 12½-gauge is 48 percent more conductive than 14½-gauge wire, and 62 percent more conductive than 16-gauge wire. The same energizer that can charge 8 miles (13 km) of 12½-gauge wire can therefore charge only 5 miles (8 km) of 14½-gauge wire or 3 miles (5 km) of 16-gauge wire.

Unfortunately, there is no rule for determining the best gauge to use based on controller size and fence length. Most fences, even relatively short ones, are made of 12½-gauge wire — not just because of its lower resistance, but also because it is easier to see, so animals are less likely to get tangled in it. Wire of still lower gauges are even more visible and more conductive, but they're less cost-effective and harder to work with.

Because current is carried along the outside of the wire, the wire's coating affects its conductivity. Aluminized wire is slightly more conductive than galvanized wire. Aluminum and zinc both slow the formation of rust, and rusty wire is far less conductive than shiny new wire. In galvanized wire, Class 3 resists rust longer than Class 1.

Plastic-coated wire is unsuitable for electric fences because plastic is an insulator — even if a pulse did travel down the wire, you couldn't feel it through the plastic coating. It can, however, be used as a neutral "sighter" wire to make an electric fence more visible to horses and humans. Coated wire for this purpose is generally 8½-gauge steel, coated to the thickness of 5-gauge wire. All-plastic wire may also be used as sighter wire. Whether plastic or plastic coated, the wire is either strung as the top wire or alternated with metal hot wires. (Electro-plastic options are discussed in the next chapter.)

Most electric fence wire is made of either steel or aluminum alloy because both are highly conductive. Steel wire is stronger than aluminum and is therefore used more often for permanent fences. The strongest and longest lasting of all steel wire is Class 3 galvanized high-tensile. Sometimes you will see barbed wire electrified, a practice that is both unnecessary and dangerous. Animals, including human ones, can get tangled in it. In their panic they may not be able to break away and can die of fright or electrocution.

Aluminum alloy wire won't rust and has four times the conductivity of steel. For a fence of equal length you can therefore use a smaller energizer, or for a given energizer you can string a longer fence. Aluminum has one-third the weight of steel, so is easier to handle. Because it weighs less and puts less tension on posts, aluminum wire requires less anchor-post bracing than steel.

While steel wire is measured according to the Standard Steel Gauge (see chapter 5), aluminum wire is sized according to the American Wire Gauge. For any given gauge, aluminum wire is smaller and weaker than steel. It therefore takes a lower gauge to withstand the same amount of stretching.

Because aluminum is softer than steel, it must be wrapped around each insulator to minimize sagging, and should be tensioned only by hand. For these reasons, aluminum is more often used for scare wires and temporary fences than for permanent ones.

No matter what wire you use, for best conductivity stick with the same kind throughout your fence. Joining two dissimilar metals (or metal coatings)

WIRE GAUGE COMPARISON

Aluminum Wire*			Approximate Steel Wire** Equivalent
GAUGE	INCHES	MM	GAUGE
9	.1144	2.9	11½
10	.1019	2.6	12
11	.0907	2.3	13
12	.0808	2.1	14
13	.0720	1.8	15
14	.0641	1.6	16
15	.0571	1.5	17
16	.0508	1.3	18

* American Wire or Brown & Sharpe Gauge
** Standard U.S. Steel Wire Gauge

causes connections to corrode more rapidly. For good conductivity, make sure all connections are firm from the start. Loose contact is an all too common problem, and one that is difficult to trace.

For a permanent fence, make all connections with one of the splicing devices described in chapter 5. If you simply tie or wrap wires together, corrosion will build up over time and reduce contact. Reduced contact means fewer joules get through but, since voltage stays the same, your voltmeter can't detect that your fence is losing its effectiveness.

Gullies

Whenever an electric fence crosses a dip or gully, make sure the bottom wire maintains its proper distance from the ground. Depending on the depth and width of the gully, you may need to add extra posts, an anchored spacer, extra wires, or a hot loop.

A hot loop is a loose wire attached to the bottom wire of the fence and positioned to follow the contours of a round-shaped gap beneath it. If the gully is fairly flat, add one or more sets of extra insulators at the bottoms of the posts on either side and string additional hot wires.

Where a section of fence crosses a flood plain or is subject to gully washers, construct anchor assemblies on both sides and splice all line wires at midpoint with wraps, knots, or single compression sleeves. Any of these splices should break if flooding causes debris to pile up against the fence. On one side, connect hot wires through an energy limiter (described in chapter 13) that shuts off the current in rising water. Energize the remainder of the fence from the other direction — if the flooded section gets washed out, the rest of the fence will still be functional.

POSTS

The size and depth of line posts for an electric fence aren't nearly as critical as for a nonelectric fence because an electric fence doesn't incur as much animal pressure. A depth of 10 to 12 inches (25 to 30 cm) is usually sufficient in heavy soil, while light or sandy soil may require a depth of 2 feet (60 cm)

or more to minimize overturning.

Fiberglass posts are becoming ever more popular for electric fences because they are self-insulating. They therefore save the trouble and cost of using insulators, and they eliminate leakage caused by dirty or defective insulators.

Traditions die hard, though, and many fencers still prefer metal and wood posts. It goes without saying that metal is an excellent conductor of electricity. Used in conjunction with a high-voltage energizer, metal posts are more likely than wood posts to cause arcing at insulators.

Dry wood is highly resistant to electrical current. Theoretically, you shouldn't need insulators with pressure-treated hardwood posts or untreated posts of ash, hickory, maple, ironbark, or red and white oak, dried to 16 percent moisture. Your electric fence will work well as long as the posts stay clean and dry.

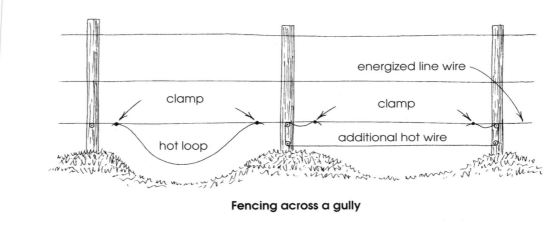

Fencing across a gully

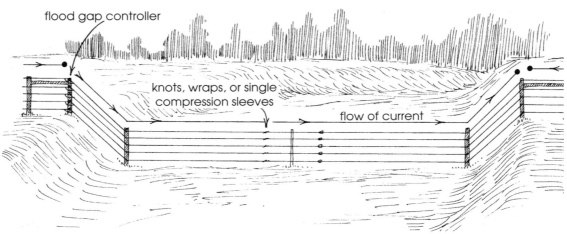

Fence crossing a flood plain

But, since moisture is a good conductor, even the most resistant wood becomes conductive in rainy or humid weather. If the post is treated with a preservative, rain dissolves the chemical salts, increasing conductivity and speeding up wire corrosion. Dust and cobwebs collect on the rough surface of wood and further increase conductivity.

If you don't use fiberglass posts, avoid shorting out your fence by attaching all pulsed wires through insulators.

INSULATORS

Even if you have the best energizer in the world and the most conductive wire, your fence will be worthless if you use inadequate insulators. Quality insulators are especially important with Class 3 galvanized high-tensile wire, since you will want the insulators to last at least as long as the wire does.

If you're building a high-tension fence, use only insulators designed for the purpose; those of lesser quality won't hold up. But even top-quality insulators won't do the job if you let them collect dirt, dead bugs, cobwebs, wet leaves, or anything else that creates a bridge between a hot wire and a post.

Insulators provide a way to fasten pulsed wire to posts while keeping the two far enough apart to prevent energy leaks. When wires go around a stout corner post, you may need two insulators, side by side. Besides guarding against leakage, insulators also reduce line wire friction at posts, and so are often used for neutral or earthed wires as well. If you want the flexibility of being able to change from hot to ground, you *must* insulate every wire.

A good insulator holds wire firmly and preserves its distance from the post, yet allows the wire to move freely in case of sudden impact. It is best, therefore, not to wrap the wire around the insulator but to let it float freely through.

When attaching an insulator to a wood post, avoid driving nails or staples so tightly that they distort the insulator. To prevent rust, which leads to rough surfaces that readily trap dirt, affix insulators with aluminum nails, stainless steel screws, Class 3 galvanized staples, or galvanized T-post clips.

Face insulators toward the *inside* of the fenced area if you are keeping livestock in, toward the *outside* if you are keeping varmints out. Eliminate

the possibility of getting your wires crossed by installing insulators when you put in the posts and attaching each line wire as you string it. Don't overlook using insulators to keep hot wires from touching anchor-post braces.

Insulators may be made of any material that does not conduct electricity, including porcelain, plastic, rubber, fiberglass, and nylon. Sizes and shapes vary, depending on the kind of post the insulator will be attached to and its position within the fence. While the insulators depicted here are representative, they by no means cover the entire spectrum.

Porcelain Insulators

Porcelain or ceramic insulators can be expensive, but good ones, treated properly, will last indefinitely. How well they hold up depends on the integrity of their glazing. If the glaze is not continuous, the porcelain will absorb moisture, lose insulating value, and crack in cold weather.

Glazing faults are not common, but cracks, chips, and scratches are. To test a batch of porcelain insulators for defects, weigh them in a bucket. Fill

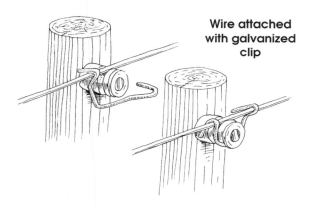

Wire attached with galvanized clip

Wire attached with short length of galvanized wire

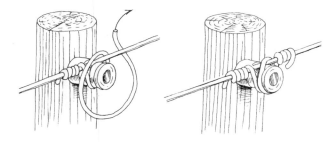

Porcelain insulators

the bucket with water, soak the insulators overnight, and weigh the insulators again. If they gain weight, don't use them.

Plastic Insulators

Because porcelain is designed only for use with wood posts, and because it tends to break if you drop or whack it during installation, most fencers prefer plastic. The best-quality plastic insulators are made of high-density polyethylene, which won't turn brittle and crack in cold weather as will polypropylene.

The most common colors are yellow, black, and red. Black insulators are the longest lasting because they are the least sensitive to deterioration in sunlight. Cheap brands, though, are made from ground-up junk plastic and are just that — junk. You are lucky if you can get one onto a post without breaking it.

Plastic insulators vary widely in design, depending on the kind of post they will be affixed to. Nail-on insulators, which can also be attached with staples, are for wood posts. Screw-on insulators are for rebar posts. Snap-on insulators are for T-posts. Some T-post insulators have a long flange, putting extra distance between post and wire to reduce the chance of arcing.

Push-in pin lock insulators are among the handiest because you don't have to thread line wire through them or bend the wire to get it into a slot. The plastic pin won't take much pressure, however, so if you're putting up a tension fence, substitute a Class 3 galvanized nail or staple.

Snap-on pin lock insulators for T-posts come in two sizes to accommodate different post widths, so be sure you get the right size. The larger ones are for posts measuring 1⅜ inches (35 mm) across the face. The smaller ones fit 1⅛-inch (28.5 mm) posts.

Tube Insulators

Some insulators come in the form of tubes, either cut to length for standard posts or sold in coils so that you can cut your own. The latter is handy for insulating wire where it rubs against a brace assembly, tree, or other potential source of leakage. Tubing can also be used to insulate lead-out and jumper wires, including those run underground. Be sure to get the right size for your wire gauge.

Position tubes so no bare wire touches a post. Staple them just tightly enough so they can't slip aside and let the wire touch. Use large staples that grip the tube without cutting into it. Some tubes have ribs or ridges so staples can't cut through. Never wrap tubing around a square post — the corners will rub and the insulator will eventually fail.

The chief disadvantage of tubes is that they have to be threaded onto a line wire before the wire is secured at both ends. Since they won't slide over a splice, slip on any you need before joining two wires together. If you're building a tension fence that goes around a corner post, make sure all tubes are on the correct side of the post before tensioning the wires. To remind you where tubes go and what length each must be, make notes on your map before you start your fence.

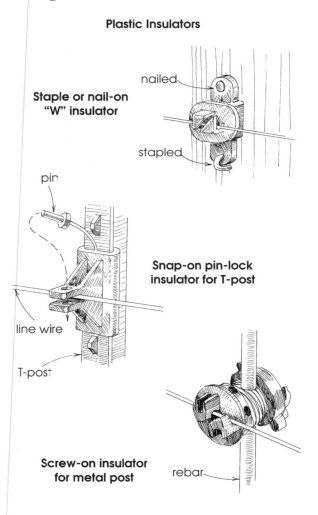

Plastic Insulators

Staple or nail-on "W" insulator

nailed

stapled

pin

line wire

T-post

Snap-on pin-lock insulator for T-post

Screw-on insulator for metal post

rebar

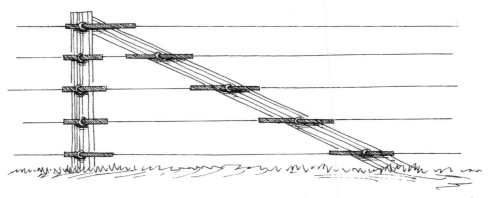

Tube insulators are threaded onto line wires before the wire is attached to posts.

Tubes are made of both polyethylene and polypropylene. The former won't get brittle and crack — a distinct advantage, since energy leakage through a tube is difficult to trace. All tube insulators tend to retain moisture, hastening line wire corrosion.

Homemade Insulators

In a pinch, recycle plastic 2-liter soft drink bottles to make insulators that won't break easily, like porcelain, and won't crack in cold weather, like cheap plastic. With a utility knife, cut off the bottle neck 1½ inches (40 mm) from the top. Fasten it to a wood post with a 16d nail, driven through a hole drilled in the center of a 1½-inch (40 mm) length of ½-inch-diameter (13 mm) PVC pipe. Rest the line wire against the bottle's collar and fasten it in place with a smooth wire tie. Alternatively, if necessary to prevent sag, pull the wire taut and wrap it around the bottle's threads.

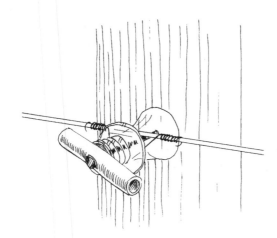

Homemade bottleneck insulator

Post-Top Insulators

Post-top insulators let you easily electrify a barrier fence or string an additional hot wire at the top of an electric fence. They are used mainly to keep horses from leaning or rubbing on a fence, and to protect horses from getting hurt on the jagged tops of posts.

Post-top insulators

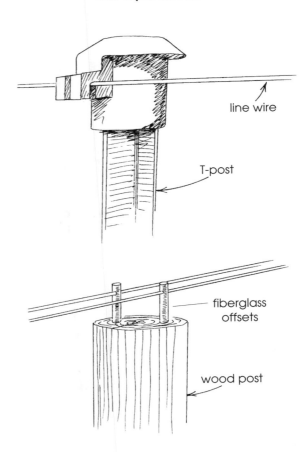

line wire

T-post

fiberglass offsets

wood post

A plastic insulator designed for use with T-posts is set on top of the post and driven down with a mallet. For wood posts, use fiberglass offsets (described later in this chapter). Drill a hole of suitable diameter in the top of the post and pound in the offset. For a wire-return system, set two arms side by side, as illustrated.

End Post Insulators

Anchor post insulators must be able to withstand full wire tension pulling in one direction. They are therefore designed differently and are stronger than line post insulators. Three basic styles work well: donut, double-U, and wrap-around.

End post insulators

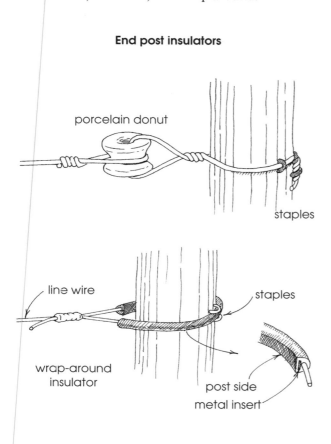

porcelain donut

staples

line wire

staples

wrap-around insulator

post side
metal insert

A porcelain donut, also called a "reel" or "O-insulator," attaches to a post by means of a short length of smooth wire through its hole. The line wire wraps around the outside of the insulator and is tied off with a standard end splice. You can also use a donut to turn a corner by attaching it to the post with wire wrapped around the outside, and threading line wire through the hole.

A double-U insulator works like the donut and comes in all manner of variations, diversely called "terminal," "end post," "strainer," "notched," or "bull-nose" insulators. Most are made of high-density plastic and most are white, although black withstands ultraviolet light better. Steel-reinforced double-U insulators are less prone to failure than porcelain donuts.

A wrap-around insulator transfers the strain of wire tension to the anchor post and has a galvanized metal insert to keep the tensioned wire from cutting through. Be sure to orient the metal insert, or flat side, toward the post. Rather than straddling the insulator with staples, drive in one horizontally above it and another below it to keep the insulator from sliding up and down the post.

When a tension fence is short enough for its in-line tensioners to be at one end near an anchor post, you can avoid the cost of end post insulators by using tensioners with built-in insulators. Don't use these in midline, though, since there you want the pulse to get through.

DESIGN

An electric fence doesn't need to be as solid as a fence designed to be a physical barrier because it doesn't have to withstand the pressure of animals leaning on it, bumping against it, or trying to get through it. It must, however, be sturdy enough to withstand initial pressure exerted by untrained stock and unwary wildlife.

A low-tension fence is cheaper and easier to build than a high-tension fence, but it's not nearly as effective because the wires can more easily be spread apart. If an animal has to force its way between the wires, it will be discouraged by shocks to the nose and ears. Shocks about the face cause an animal to pull back. Shocks elsewhere on its body cause it to jump forward and get through.

When you string an electrified tension fence, you can prevent digging and crawling by creating a physical barrier at the bottom. Simply string a few closely spaced neutral or earthed wires and attach them to additional short line posts set 6 to 20 feet (2 to 6 m) apart.

The foundation for a low-tension fence need only be strong enough for you to stretch the wires sufficiently tight to prevent sagging. Set a stout wooden post at least every 180 feet (55 m), set a steel T-post every 30 to 40 feet (10 to 12 m) in between, and add fiberglass spacers as needed to reduce sag and maintain wire spacing where the fence changes direction or elevation. The more wires the fence has and the more closely they're spaced, the more line posts and spacers you'll need.

For either a low-tension fence or a high-tension fence with up to six wires, single-span anchor assemblies work well. You will need double-span assemblies for a tension fence with seven wires or more. Line posts for a tension fence on level terrain can be as far apart as 100 feet (30 m), with spacers at the midpoint. In hilly country, you may need to set posts as close together as 15 feet (4.5 m).

Construction of an electrified tension fence is similar to the construction of a nonelectrified one. The chief difference is that the electric fence has fewer wires under considerably less tension. Compared to wire tension between 250 and 300 pounds (1,130 to 1,350 N) for a barrier fence, the electrified version needs only 100 pounds (450 N), with 150 to 200 (680 to 900 N) being standard.

The number of wires, their spacing, and the distance of the bottom wire from the ground are determined by the specific animal to be controlled. Variables include whether animals of different species are kept together; whether they are young, mature, or both; and whether predators must also be deterred.

Since most electric fences offer little by way of a physical barrier, some animals, regardless of species, simply will not be deterred. Difficult-to-control animals are usually those with strong motivation to get in or out and no early training to respect an electric fence.

Wild animals are more aggressive than domestic ones and therefore harder to control. A predator-proof fence must be too high for them to jump over, closely enough spaced so they can't push through, and low enough so they can't dig or crawl under. Where dogs and their kin are involved, the fence must also carry a strong electrical charge. Canines lack sweat glands, so they have little moisture and salt on their skin to improve electrical conductivity.

Even though wire is the cheapest part of an electric fence, stringing too few wires is a common error. The addition of just one more wire will often double a fence's effectiveness yet add minimally to its cost.

Unfortunately, no hard-and-fast rules have been established for deciding exactly how many line wires you need and how far apart they should be. The possible combinations of climate, terrain, livestock type and temperament, and predation problems are virtually endless. The same fence that works perfectly for one person may require minor or major adjustments for a seemingly identical situation elsewhere.

Happily, experienced electric fencers have developed a few good guidelines. For starters, the most important pulsed wire is the one at nose height for the animal being controlled, or about two-thirds the animal's total height. If a herd contains animals of more than one species or size, you will need more than one hot wire. The bigger the animals, the taller the fence must be and the more line wires it will require.

When animals of different ages, sizes, or temperaments are penned together, or if more than one kind of muncher is pillaging your garden, the fence must be designed for the most difficult-to-control among them. By the same token, boundary fences generally have more line wires than cross fences because an animal breaking out, or a predator breaking in, is usually a far more serious problem than the occasional animal wandering onto the wrong pasture.

Guidelines

An electric fence works only as long as the pain is greater than the gain. Pain, of course, refers to the strength of the fence's zap. Gain refers to the animal's motivation, be it food, water, or the desire to breed. The greater the motivation, the better your fence must be. Since motivations ebb and flow with the season, design your fence for those times when they're strongest.

The easiest animals to confine are dairy cows with no calves and plenty of good pasture. In areas of even rainfall, these legendary contented cows can be controlled with a single electrified wire. Depending on the size of the cows, string the wire between 24 and 36 inches (60 to 90 cm) above the ground. In drier areas, you will need a second, grounded wire, 4 inches (10 cm) below the first.

Add a third, hot wire toward the bottom to contain cows with calves, spacing it about halfway between the second wire and the ground. The same fence will hold most beef cattle, with or without calves. To confine bulls, you may need as many as four strands, the top one about 45 inches (115 cm) high.

Pigs are relatively easy to control with electrified wire; it deters them from rooting or pushing beneath the fence. You shouldn't have trouble keeping hogs in with a three-strand wire-return system, provided the lowest wire is no more than 6 inches (15 cm) from the ground.

Sheep and goats require more strands than cattle or hogs because they are more difficult to confine and because they are of greater attraction to predators. Woolly sheep are particularly hard to shock, except after shearing or when they're wet. In addition, sheep learn slowly and forget easily. Goats learn quickly and don't soon forget, but they do like to jump, as do some breeds of sheep. A five- or six-strand wire-return system, with a total height of about 40 inches (100 cm), will control most sheep and goats with young.

A six-strand wire-return system with spacers every 20 feet (6 m) keeps marauding foxes out of your henhouse, wild pigs out of your garden, and — should they be a problem in your neck of the woods — kangaroos and dingoes out of your pasture. This fence will also repel casual coyote and dog incursions.

Where dogs and coyotes are a serious threat, you will do better with an eight- or nine-strand wire-return system having a total height of about 45 inches (115 cm). Since canines search out gaps, keep the bottom wire tight, hot, and following the contour of the land at no more than 6 inches (15 cm) above the ground.

Horses, especially those with shoes, are particularly sensitive to shock, making an electric fence an inexpensive alternative to a rail fence. Nine wires will control most mares and stallions. Electrify the bottom wire to keep them from pushing the fence; electrify the top wire to discourage them from leaning over and weighing it down. For mares with foals, add one to three wires toward the bottom so the foals can't roll under and dogs can't crawl in.

A ten-strand wire-return tension fence will keep most livestock in and most predators out, and doubles as an effective physical barrier should the power go out.

The accompanying chart offers a starting place for designing your electric fence. The spacings can be followed exactly if you use wood or fiberglass posts. If you use steel T-posts you will have to make minor adjustments according to how the insulators fit between the lugs.

GARDEN FENCES

A single wire, 10 to 12 inches (25 to 30 cm) high, will keep small dogs from digging in your garden. If salad-seeking rabbits are your problem, a single strand 3 to 4 inches (7 to 10 cm) high should do the trick. To deter raccoons from your corn patch, string two strands spaced 5-11 (13-28 cm). Space posts 6 feet (2 m) apart.

An all-purpose three-wire fence, spaced 4-5-6 (10-13-15 cm), should deter most small dogs, rabbits, raccoons, woodchucks (groundhogs), and cats. If any slip through, try 3-4-6-8 (8-10-15-20 cm). These fences are so short you don't need a gate. Simply step over — but if you like to garden barefoot, watch your toes.

Deer and Elk Fence

Deer and elk offer gardeners a special challenge. No inexpensive, surefire way has yet been found to keep them out. White-tailed deer, for example, can belly under a 6-inch-high (15 cm) wire or jump over a 5-foot (1.5 m) fence from a standing start. Innocent-looking wires, however, lead them to believe they can slip through, and the resulting jolt to the nose offers sufficient discouragement.

The size of the deer, their population, and whether you have more than one species to contend with all make a difference in your fence design. The fence that works for you will also depend on how motivated the population is. Strong motivations include a scarcity of food and water everywhere except inside your fenced area, or a fence that blocks a path of migration, or one that excludes does from their usual fawning grounds.

The size of the enclosed area is also significant. Deer are less inclined to jump into a small space than into a larger one. Permanent plantings such as

ELECTRIFIED WIRE AND POST SPACING

Animals (Purpose)	Number of Wires	Wire Spacing*		Post Spacing**					
				LEVEL		ROLLING		STEEP	
		IN	CM	FT	M	FT	M	FT	M
Dairy cows	1	**30**	76	90	25	65	20	50	15
Beef cattle	2	**20**-16	51-41	90	25	65	20	50	15
Raccoons & small animals	3	**4-4-5**	10-10-13	6	2	6	2	5	1.5
Poultry	3	7-7-7	18-18-18	30	9	30	9	25	8
Hogs	3	**6**-7-**8**	15-18-20	50	15	50	15	35	10
Cows & cattle w/calfs	3	13-11-**12**	33-28-30	90	25	65	20	50	15
Bulls, difficult cattle	3	17-9-**11**	43-23-28	50	15	50	15	35	10
Bears	4	4-5-**6**-20	10-13-15-51	20	6	20	6	15	5
Sheep	4	6-7-10-**12**	15-18-25-30	30	9	30	9	25	8
Sheep w/lambs	5	6-7-**8**-9-**10**	15-18-**20**-23-**25**	30	9	30	9	25	8
Buffalo & bison	5	**13-13-13-13-13**	33-33-33-33-33	45	14	45	14	14	4
White-tailed deer	5	**10-12-12-12-12**	25-30-30-30-30	50	15	50	15	35	10
Fox & wild pig	6	2-**6**-**6**-**8**-**8**-9	5-**15**-15-**20**-20-23	30	9	30	9	25	8
Goats	6	**5-5-6-7-8-9**	13-13-15-18-20-23	50	15	50	15	35	10
Mule deer	6	8-**10**-10-**10**-10-**10**	20-**25**-25-**25**-25-25	50	15	50	15	35	10
Predator control	8	4-**5**-5-**5**-6-6-7-**8**	10-**13**-13-**13**-15-15-18-**20**	30	9	30	9	25	8
Mares & stallions	9	16-**5-5-5-5-5-5-5-6**	41-**13-13-13-13-13-13-13-15**	50	15	50	15	35	10
Elk & antelope	9	10-**10**-8-**8**-8-**8**-5-**8**-12	25-**25**-20-**20**-20-**20**-13-**20**-30	50	15	50	15	35	10
All livestock, most predators	10	4-4-4-**4**-5-**5**-5-**5**-5-**6**	10-10-10-**10**-13-**13**-13-**13**-13-**15**	50	15	50	15	35	10
Mares w/foals	10	12-**5-5-5-5-5-5-5-5-6**	30-**13-13-13-13-13-13-13-13-15**	50	15	50	15	35	10
Security fence	15	2-**6-6**-6-**6-6**-6-**6-6**-8-**8-8**-8-**8-8**	5-**15-15**-15-**15-15**-15-**15-15**-20-**20-20**-20-**20-20**	50	15	50	15	35	10

* wires shown in **bold** should be energized, the remainder earthed.
** add one or two spacers where posts are more than 20 feet apart.

berries or fruit trees are more attractive than short-rotation crops, especially if nibblers have already begun making regular visits. Any fence works best if it is put up *before* the pilfering starts.

Your fence design will depend, too, on whether you must keep out *all* deer or only most of them. You don't have to fence in the whole area if occasional nibbling is acceptable. Suppose you are growing a large field of corn, and deer tend to approach from a woodlot bordering one or two sides. You can reduce damage considerably by running a fence along the tree line.

Any fence will be more effective if it can be seen from a distance so deer will approach with caution. Keep 5 to 10 feet (1.5 to 3 m) cleared of brush around the outside edge. To make the fence even more visible, tie on flags of aluminum foil. Curious creatures coming to investigate will get zapped and back off.

The five-wire deer fence designed at Pennsylvania State University and listed in the accompanying chart works well for white-tailed deer. To deter mule deer and elk, try the six-wire fence. For serious incursions of elk and antelope, you may need nine wires. Place line posts no more than 60 feet (18 m) apart and add spacers every 20 to 30 feet (6 to 9 m). To keep out raccoons, woodchucks, and rabbits as well, include a few additional lower wires.

Worth mentioning, if only to help you avoid getting overenthused if you see them described elsewhere, are two-dimensional and three-dimensional deer fences. The two-dimensional deer fence, also called a "modified figure 4," "Kiwi," or "New Hampshire" fence, is actually two parallel fences spaced 3 feet (90 cm) apart, one side hot, the other grounded. A three-dimensional fence is a complex rig consisting of a fence built so it slants toward the deer side. Both these fences are more difficult to construct than a standard vertical fence and both make weed control difficult (if you clear manually or mechanically) or expensive (if you spray).

Tree Protection

To protect a row of newly planted trees from deer, place a stout anchor post at each end. Space pairs of braced line posts every 65 feet (20 m), crossed at ground level, as illustrated. Set them at an angle of 60° to each other and 60° to the ground. String three hot wires on each side, spaced 15-12-12 (38-30-30 cm), run from one end post to the other. To keep rabbits from chewing on the bark, add a fourth wire, 4 inches (10 cm) from the ground.

BEAR FENCE

With landfills closing right and left, bears are more likely than ever to snoop in backyard garbage cans. Bears are also a problem for beekeepers, who don't think it's so funny "how a bear likes hunny." An electric fence offers a fast, inexpensive deterrent, provided it doesn't cross the bear's traditional runway — otherwise you're fighting a losing battle.

A single hot wire, 20 inches (50 cm) off the ground on posts 15 to 20 feet (4.5 to 6 m) apart, deters bears fairly successfully. If one wire doesn't stop your bear, try three, spaced 10-10-10 (25-25-25

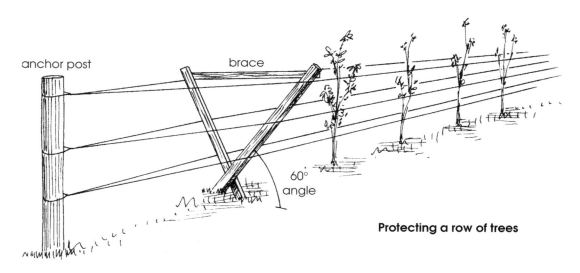

anchor post   brace

60° angle

Protecting a row of trees

cm). If three wires still won't do it, put up the four-wire bear fence listed in the chart.

Tie strips of fresh bacon or ham rind at intervals along the fence to attract bears and encourage them to touch the wire with nose or tongue. String the wires at least 3 feet (1 m) away from the garbage can or hives so the bears can't foil you by reaching through.

If you keep bees in a remote spot, protect your battery energizer from ursine or human vandals by hiding it in a deep hive body with holes drilled in the side for the lead-out wires. Set this body on top of a working hive, separated from it by an inner cover with the hole taped shut.

As a backup in case your fence gets shorted out or the battery goes dead, rig a freon-powered boat horn to trip wires inside the fenced area. If a bear does get in and starts tearing things up, it will trigger loud blasts on the horn. Even when no one is close enough to come to the rescue, the noise may be enough to scare off the intruder.

SCARE WIRES

The chief disadvantage of an electric fence is that it must be hot at all times to be effective. It therefore makes sense, especially in a remote area, to build a fence that doubles as both a physical and a psychological barrier. You can easily do so by constructing a nonelectrified fence and augmenting it with scare wires.

A scare wire, also called an "offset" or "stand-off" wire, is an energized wire strung parallel to a fence and 6 to 8 inches (15 to 20 cm) out from it. If the fence is made of wire, and is well grounded, any animal touching both the fence and the scare wire will complete the circuit. Take care the scare wire doesn't touch the fence, though, and energize the whole thing. An offset for a wood or plastic fence relies on ground return.

A scare wire can be added to any existing non-electric fence to make it predator-proof and to increase its longevity by keeping livestock away. If you are replacing an old fence and can't afford to do it all at once, a scare wire will keep it functional while you replace one section at a time. A scare wire can double as a feeder line to get electricity from one fence to another.

Where you place the offset depends on its purpose. If you want to keep stock off the fence, run the wire along the *inside*. To keep animals from digging out, string it 8 to 12 inches (20 to 30 cm) above the ground. To keep animals from leaning or rubbing against the fence, string the wire at two-thirds of the animal's height. If livestock lean on the fence top or predators jump over, string the wire along the top.

Most predators belly under a fence or try to climb through. For good control of wild pigs, dogs, coyotes, and wolves, string a scare wire 8 to 12 inches (20 to 30 cm) aboveground on the *outside* of the fence. Any animal trying to crawl or dig under will get the wire across its back. Any animal trying to climb through will get it across the belly.

If you have more than one purpose for adding scare wires, string them top and bottom, inside and outside. But there comes a point of diminishing returns, when the cost of adding scare wires is greater than the cost of constructing a good electric fence.

The chief expense is in the insulators, also called "brackets," "arms," or "isolators," used to attach the scare wires. Arms of fiberglass or plastic attach to posts. Wire brackets attach to woven or stranded wire, but should be placed near a post for best support.

Space brackets no farther apart than 65 feet (20 m), or as close together as necessary to keep the scare wire from sagging. When adding scare wires to an old fence, watch for frayed ends that may touch the hot wire and short it out.

Homemade Offsets

You can make low-cost offsets for wood posts by cutting ¾-inch (19 mm) PVC pipe into 8-inch (20 cm) lengths. Cut a ¼-inch-deep (6 mm) slit into one end of each and drill two holes through the other end, as illustrated. Fasten each pipe to a post with two 14d common nails, driven through the two holes. Slip the scare wire into the slit and insert a nail into the pipe to hold it there. These offsets should last at least five years.

TRAINING

Animals, both domestic and wild, can easily go under, over, or through most electric fences if they

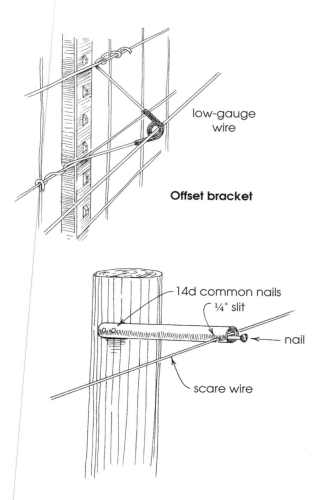

low-gauge wire

Offset bracket

14d common nails
¼" slit

nail

scare wire

Homemade offset

really want to, so they must be taught respect. Training entails making sure the animal gets its initial shock on a sensitive part — tongue, nose, or ear — which is likely if the animal approaches the fence out of cautious curiosity.

A jolt about the face causes an animal to back off. If the animal first encounters the fence at a dead run, it may feel little or nothing through thick fur. Even if the animal gets zapped partway through, its natural tendency will be to move forward. Once the animal gets through, it may never develop respect for electric fences.

Training wildlife requires keeping the fence line cleared so creatures can see the wires from a distance. You might also bait hot wires with food or shiny objects to encourage curious individuals to investigate. Despite your best efforts, wildlife may still get through when being chased by dogs.

Domestic animals that grow up with electric fences learn to respect them from the start. Older animals introduced to electric fencing for the first time require a period of training. Never turn animals loose in a new yard, or they may bolt. And never get them to touch a wire by driving them into it. It is best not to be nearby for that first encounter, or your animals may associate *you* with the pain.

To train inherently spooky or skittish stock, confine them to a small electrified enclosure. Make sure the number and strength of the electrified wires offer a physical barrier as well as a psychological one, or run scare wires along a sturdy barrier fence. If necessary to ensure maximum shock value in the training fence, temporarily disconnect other fences attached to the same energizer.

The best training occurs during a time of minimum stress, so leave the animals alone to explore at their leisure. Curious animals such as goats will investigate on their own. To encourage hogs, cattle, horses, and mules to touch or reach through the fence, tie on an ear of corn or bag of hay, attach a tin can or other shiny object, or place a bit of feed on the other side.

Average training time ranges from twelve to forty-eight hours, depending on the number of animals and their temperaments. When animals thus trained are turned into their area of permanent residence, they will soon locate the electric fence and graze to within a few inches of it. If you then remove the fence, they will often stick to the same graze line, at least for a while.

Two exceptions are horses and humans who don't pay attention to where they're going. For their benefit, make a hot fence more visible by flagging it with colorful strips of plastic ribbon (the sort used by surveyors), by making the top wire a wide band of electrified ribbon (described in the next chapter), or by adding plastic sighter wires (described earlier in this chapter). The thinner the line wires, the harder they are to see and the more important it becomes for you to render the fence visible.

Livestock and predators alike constantly test an electric fence, reaffirming their fear with each new jolt. Predators are especially quick to take advantage of a breach, so it's important to maintain the pulse at deterrent level. Well-trained livestock tend to be a bit slower on the uptake, often taking days or even weeks before discovering freedom.

CHAPTER 10 ▼▲▼▲▼▲▼▲▼▲▼

TEMPORARY FENCES

▼▲▼▲▼▲▼▲▼▲▼▲▼▲▼▲▼▲▼▲▼▲▼▲▼

Temporary fences have so many uses that it's impossible to list them all. They are ideal for controlled grazing, where livestock have access to a limited amount of forage at any one time. They are perfect for protecting hay or silage from pillaging deer or livestock. They let you graze stock on unfenced fields or lawns. They let you easily isolate ornery or unhealthy animals and those returning from show or newly acquired. They will keep seasonal marauders out of your garden or orchard.

With a temporary system you can fence off occasionally hazardous areas such as an autumn woodlot where oak or cherry trees grow or a boggy place where stock may mire. You can temporarily fence a pasture while putting in permanent fencing, or can set up night paddocks on trail rides. The possibilities are almost limitless.

ELECTRIFIED OPTIONS

Temporary fencing should be easy to put up, easy to take down without damage, and effective for the job at hand. For livestock or predator control, electrified versions are the best options because they are lightweight and don't require heavy posts or specialized tools.

A temporary electric fence works essentially like a permanent one. The temporary version, however, uses posts that can be pushed into the ground by hand or set with a mallet. Sometimes rebar posts with insulators are used. More often the posts are self-insulating plastic or fiberglass. Guys are used to stabilize corner posts.

For the sake of convenience, the energizer is usually battery operated, but a battery drains quickly unless you move the fence often enough to avoid weed overload. Where electroplastic line wire is involved, use only a low-impedance energizer with an on-time no longer than 0.003 seconds, or you will run the risk of melting the plastic. Even though the fence is temporary, don't forget a ground rod.

Temporary fences are usually energized and earthed by means of alligator clips or by simple wraps. If you are linking the temporary fence with a permanent one, electrify it with a curl-on connector or a fence "isolator." The latter consists of a tail you wrap around the temporary wire and an insulated handle surrounding jaws that grip the permanent hot wire.

REEL FENCES

If you are stringing a parallel wire fence, wind the wire on reels so it won't get tangled or knotted when you move the fence. Reels come in varying sizes to accommodate different wire lengths. Some reels have a ratchet brake to stop the spool from turning when you stop pulling wire. Some have a lock so the wire can't unwind on its own. A good reel has a carrying handle, adjustable drag so the wire won't run away from you, and a crank for rewinding. To string a reel fence, attach the wire to one end post and let it feed out as you carry the reel to the other end post.

Among reel accessories are posts designed to support one or more reels. Set the post at the end of the fence line, adjust the reel for a slight drag, and pull out only as much line wire as you need. To save time when you are stringing a fence with more than one wire, tie the ends of each wire to a line post and let the reels all pay out at once as you carry the post down the fence line.

Another accessory is a multiple wind-up system that lets you rewind more than one reel at a time. The hand-cranked version doubles as a reel support post. Another version, developed by Premier Fencing, has a ground-driven wheel that automatically winds one or more reels as you push the device along, saving sore muscles in your cranking arm.

If you're on a tight budget, use reels designed for extension cords and sold at any discount department store. Besides being considerably cheaper, these reels are lighter and easier to handle. At the other end of the cost spectrum is a reel with a battery-powered energizer hidden in its hub.

The cost of a reel fence is relatively low, varying with the fence's length, the number of reels you use, and the kind of wire you string. Since standard fence wire kinks rather easily, the wire used for a temporary fence is more often aluminum, electroplastic, or lightweight steel cable.

Reels

Stringing single line wire

Reel support post

Electroplastic Twine

Electroplastic twine, sometimes called "polywire" or "electracord," has been around since the 1970s. It consists of several polyethylene strands twisted together with a few thin stainless steel filaments. The diameter and number of the steel strands vary from brand to brand. Some brands have as few as three, others as many as nine. Although six is the industry standard, nine strands give you 50 percent more conductivity.

Polywire is lightweight, inexpensive, easy to use, and can be cut with knife or scissors. Its disadvantage in cold climates is that it stretches and sags when coated with ice or snow, and in all areas it has relatively low conductivity, since only the metal strands carry current.

Manufacturers have tried to improve conductivity, either by increasing the number of conductive filaments or by improving their quality. Some brands use tinned copper instead of stainless steel; others use aluminum alloy. Both break more easily than steel.

If the metal strands break, the twine loses all conductivity. Because the plastic part remains intact, you can't readily tell that the wire has been rendered powerless. Some brands have filaments of both stainless steel and aluminum alloy. Aluminum strands offer greater conductivity until they break, then the stainless steel strands take over.

The number of twists per inch also varies with the brand. The looser the twist, the more easily the twine stretches. Stretching causes non-stretchable conducting filaments to break, so the tighter the twist, the better the twine. In some brands, the metal filaments are crimped to minimize breakage.

The number and color of plastic strands vary as well. Since most animals can't readily distinguish color, whether the twine is yellow, white, orange, or black affects its visibility only for humans. Color does, however, affect resistance to ultraviolet sunlight. Black twine is the most impervious, but it is the hardest for humans to see and therefore the least popular. The most popular color is orange. For a wire-return fence, installation and troubleshooting are both easier if you use one color for hot wires and another for grounded ones.

Electroplastic twine is manufactured worldwide, accounting for the tremendous number of variations. Talking about how long it will last is therefore a bit like talking about Texas — anything you say is true. Some polywire lasts up to ten years; a reasonable average is four or five. Longevity, of course, is influenced by how roughly you treat the wire and whether you store it inside for part of the time or leave it out in the elements year-round.

Because electroplastic twine is easy to tie, most people splice it with knots. But knotting reduces its conductivity by at least half because air gaps minimize contact between the conducting filaments. A Western Union splice (see chapter 5) both improves contact and makes a stronger joint.

Since polywire's fine filaments are highly resistant to electrical flow, this wire should be used only for short runs of fence that aren't overgrown with weeds. Standard six-strand twine conducts well for approximately 2,000 feet (600 m). Even so, to overcome resistance, you will need an energizer that puts out at least 1,000 volts more than you would need for a permanent fence of the same length.

Soft Steel Cable

Galvanized steel cable, widely used in Europe, was first brought here from England under the trade name Maxishock. Consisting of fine strands of low-tension wire twisted together, it costs about the same as polywire, has eighty times the conductivity, and lasts considerably longer — from ten to fifteen years — but is not quite as kink resistant and is more prone to sagging.

Like polywire, steel cable is soft enough to knot, but you will get better connections with Western Union splices. Because this wire has low elasticity, add lightweight compression springs every ¼ mile (400 m) for good tension on long runs.

Other soft steel cable is now available that is stronger and more conductive, but it can't be tied as easily and so requires the additional expense of splicing devices, described in chapter 5.

Electroplastic Tape

Electroplastic twine and soft steel cable are both hard to see, so neither is suitable for places where good visibility is important. Since temporary fences are designed to be moved, it is easy to forget where you've put them. Horses have a hard time seeing wire, and deer have to see a fence to be deterred.

Electroplastic tape was invented to solve these problems. It is made of the same materials as electroplastic twine, but woven in a flat strip that looks like ribbon. It is also called "hot ribbon," "polytape," and "hot tape." Dual-track tape has a built-in wire-return system for use in dry conditions or on packed snow.

You can build an entire temporary fence with hot tape or you can string a single strand to render an existing fence more visible. Special insulators are designed to accommodate its width.

Hot tape can be easily seen, not only because of its width, but also because it flutters in the slightest breeze. Fluttering makes the tape visible for a greater distance by making it look wider than it is. Fluttering also interferes with an animal's depth perception, encouraging marauders to steer clear. Unfortunately, wind abrasion tends to wear the tape out rather quickly.

Hot tape comes in white, yellow, orange, and striped. White and yellow are more visible in summertime; orange is more visible against winter snows. Striped tape therefore offers all-season visibility.

Tape comes in at least two widths: ½-inch (13 mm) and 1½-inch (36 mm). Wider tape can be stretched tighter, is more visible, has better conductivity, and lasts a little longer than narrower tape. You can expect tape to last from two to ten years, depending on how much wind abrasion and ultraviolet radiation it's subjected to. Prolong useful life by taking care not to drag the tape along the ground, pull on it, or snap it. Hot tape is not as easy to wind and unwind as twine or cable, so have patience.

Like twine, tape loses conductivity when spliced with knots. To improve contact, wrap two ends around each other several times rather than knotting them. As an alternative, use stainless steel buckles designed for this purpose, or add a strip of aluminum foil to each knot. Besides improving contact, foil makes splices more visible and easier to find. Start a square knot, lay a ½-inch (13 mm) strip of foil against it, wrap the foil around the hot tape three or four times, finish tying the knot, and squeeze it tightly.

Wire Spacing

A reel fence used for perimeter fencing requires more line wires and more closely spaced posts than one used for cross fencing. Cross fencing — to divide a field that has a permanent perimeter fence — is the more common use. A cross fence usually consists of hot wires fastened by means of end post insulators to line posts opposite each other in the perimeter fence. For a temporary perimeter fence or cross fence built without benefit of permanent posts, use well-driven steel T-posts or stout fiberglass rods at ends and corners. To prevent sag, add lightweight line posts every 40 to 50 feet (12 to 15 m) along the wire.

A single strand strung at two-thirds of the animals' height is usually sufficient for horses, cattle, and pigs. For sheep you will need three wires and for goats four, spaced as shown in the accompanying chart.

Seasonal Garden Fence

For seasonal protection of gardens, orchards, or nursery plantings, build a temporary electric fence following the same guidelines as for a permanent garden fence, outlined in chapter 9. Use hot tape and you will run less risk of bumping into the fence while planting or hoeing.

Where deer or elk are the pilferers, many gardeners create a successful deterrent with a single strand of ½-inch (13 mm) hot tape, strung on posts spaced 6 to 15 feet (180 to 460 cm) apart, approximately at nose height of the species involved — that is 3 feet (90 cm) or less for most deer, about 4 feet (120 cm) for elk. Where more than one species is involved, string a tape for the larger animal and a second tape 6 to 8 inches (15 to 20 cm) below it.

Should deer still get through, run two strands. For particularly persistent pests, try four to six strands, 10 to 12 inches (25 to 30 cm) apart. If your garden adjoins a woodlot, clear a buffer strip so deer see the fence more easily and approach it with greater caution.

Deer can be especially persistent where the population is high or nibblers have already tasted forbidden fruit. To encourage respect for your fence, bait it with peanut butter. Smear some on heavy foil strips and hang them over the hot tape.

CROSS-FENCE WIRE SPACING

Animal	No. Strands	Spacing
Pigs	1	7" (18 cm)
Hogs (large)	1	15" (38 cm)
Cows	1	30" (75 cm)
Horses, deer	1*	32" (81 cm)
Sows	2	12-24" (30-61 cm)
Calves	2	20-34" (50-86 cm)
Sheep	3	8-16-32" (20-40-80 cm)
Difficult cattle, bulls	3	17-10-10" (43-25-25 cm)
Goats	4	8-14-24-36" (20-36-60-90 cm)
Deer	4	10-10-10-10" (25-25-25-25 cm)

*Hot tape

Curious deer will be attracted to the fluttering strips, try for a taste, and get a memorable jolt. The smaller the area to be protected, the better the peanut butter fence works. Effectiveness approaches 100 percent for a 50' x 100' (15 x 30 m) garden.

ELECTROPLASTIC NET

An electroplastic net fence is more expensive than a reel fence and not nearly as easy to set up. It is more reliable, however, especially for boundaries, where it does a better job of controlling predators or confining livestock during high-stress times such as breeding, weaning, and introducing new stock. Because it is easy to see, animals virtually train themselves to respect it.

This fence is woven from electroplastic twine and comes completely pre-assembled with plastic posts and hot and ground connector clips. You don't need a gate — just switch off the energizer and roll back the last panel. Support ends and corners with a guy line secured with a tent peg (guys and pegs come with the fence). If you don't like the idea

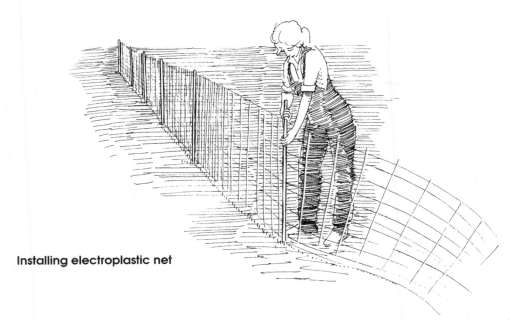

Installing electroplastic net

of guys outside the fence line, strengthen ends and corners by using 1-inch (25 mm) fiberglass rods driven 18 inches (45 cm) or deeper.

Electroplastic net was invented in the early 1960s by John Gilbert, a clever Irishman living in England. The original version, called Flexinet, had string vertical stays welded to horizontal strands of twine with tiny blobs of plastic.

Gilbert later modified the design and called it "Livestok" (sold in the U.S. as Electronet). To reduce sagging and tangling, he replaced the string verticals with semi-rigid plastic stays. To improve the fence's barrier value, Gilbert increased the number of horizontals from six to eight. To increase its useful life, he replaced the orange plastic twine with ultraviolet resistant black twine.

A third and later version, called Fast Fence, has string verticals like Flexinet, but they are knotted instead of welded, making the fence somewhat more delicate. On the other hand, knotted joints are easier to repair than plastic ones.

The cost of electroplastic netting varies with brand and height. Each brand comes in three heights. In general, the taller the fence, the larger the mesh and the less suitable it is for controlling small animals. The tallest, in the 40-inch (100 cm) range, works best for confining goats and deterring foxes, coyotes, dogs, and cougar.

The middle height, just over 30 inches (75 cm), is used to confine sheep, cattle, pigs, horses, and non-flying ducks and geese, and to keep out cats, rabbits, skunks, and groundhogs. The shortest, 20-plus inches (50 cm), is low enough to step over and can be used to confine weeder geese, small dogs, lambs, piglets, and turkeys; to protect gardens from small pests; and to control raccoons, groundhogs, and mink.

The standard version has all hot horizontal wires except for the bottom one. A wire-return version, called Pos-Neg, has alternating hot and earthed horizontals. It is used in dry climates and for confining climbing creatures such as monkeys.

The longest you can expect one of these fences to last is ten years. Black net, because it is less sensitive to ultraviolet light, lasts longest; orange net generally only half as long. Durability varies with how much ultraviolet light the fence is exposed to and how carefully you treat it.

You will get more use out of your fence if you store it away seasonally and if you don't let it drag along the ground while putting it up or taking it down. When the fence isn't in use, wrap it in plastic, tape it tightly, and hang it from a nail, away from gnawing rodents.

Installation

One roll of electroplastic net weighs about 10 pounds (4.5 kg), covers 150 feet (45 m), and can be set up in minutes by one person. Begin by mowing the fence line. Untie a roll of fencing, locate one end post, and poke it into the ground. Walking backwards, carry the rest of the roll along the fence line and drop each post as it pulls loose from your hands. Go back to the beginning and stretch each panel just tight as you insert each line post into the ground. When you join two rolls, attach the hot and ground clips in the right order or you will cause a dead short.

If your soil is hard-packed, this fence is easiest to install when the ground is moist so the steel spikes at the bottom of the posts slide right in. If the ground is dry or the spikes are rusty and you have trouble getting them in, make a pilot hole by pounding an old screwdriver or a short length of 6-gauge wire into the ground. To help rusty tips slide more smoothly, wire-brush and spray them with a light oil such as WD-40 each time you move the fence or store it for the season.

Electroplastic net sags rather easily, sometimes causing shorts. You can minimize the problem by adding three or four more posts — available from the same source as the fence — and spacing all the posts closer together.

Electroplastic net doesn't stand well in sandy or loose soil and may topple if wind blows tumbleweeds or other vegetation against it. You can solve this problem by using longer, thicker posts. Mention your concern to the supplier when you purchase the fence, and maybe you can work out a deal.

Even though it is designed as a movable fence, electroplastic netting doesn't work as well with a battery energizer as with a plug-in. Grass and weeds drain energy from a battery-powered fence, rendering it less effective.

Electroplastic netting must be kept hot at all times or rabbits may chew through it and livestock and wildlife will definitely get tangled in it, possibly with dire consequences. In lucky cases where death doesn't result, posts will be pulled out and the net may be torn. Because tears are common, a repair kit comes with each fence.

SNOW FENCE

During the past three decades, substantial research has been done on fences for snow control. Applications include protecting driveways from slush and ice, keeping drifts away from houses and barns, and trapping snow in ponds, where it later melts to provide drinking water for livestock.

A snow fence can be used effectively only in an open, windswept area where snow stays on the ground thirty days or more and wind speeds reach at least 9 miles per hour (14.5 km/h) — conditions found on over 40 percent of the earth's surface.

Most research in the U.S. is conducted at the Rocky Mountain Forest and Range Experiment Station in Laramie, Wyoming, where every kind of fence ever invented for snow management has been tested. Excluding a living fence (consisting of trees, shrubs, or grasses), the three most common kinds are plastic snow fence, vertical slat or "Canadian" fence, and horizontal board or "Wyoming" fence.

Plastic Snow Fence

Numerous designs have come and gone since plastic snow fence was introduced in the 1960s. The idea hasn't caught on yet, largely because early plastic mesh wasn't durable enough to hold up against sun, wind, and cold. Some materials tear loose in strong wind or get brittle from ultraviolet rays and cold temperatures. Others stretch so much they need an unreasonable amount of support to hold them up under heavy snow loads. Still others attract animals and either get eaten up or torn down by critters rubbing against them.

Some materials simply don't trap snow efficiently. Efficient collection requires a porosity of 50 percent. Since porosity measures the proportion of open spaces to solid fence material, a proper snow fence with 50 percent porosity looks as though it is half full of holes.

The snow fence framework developed at Laramie consists of a series of steel T-posts spaced 7 feet (210 cm) apart. Top and bottom rails are lashed to the posts with smooth wire. The rails consist of 8-foot-long (240 cm) 2x2s (5x5 cm), nailed together with a 6-inch (20 cm) overlap at each end. Wood posts can be used instead of steel, spaced 8 feet (240 cm) apart with the rails nailed directly to them.

The distance between the rails should equal the width of the fencing material, either 4 or 6 feet (120 or 180 cm). Stretch the material over the framework and sandwich it between strips of lath fastened every 16 inches (40 cm) with drywall nails.

Leave a 6-inch (15 cm) gap between the bottom rail and the ground to increase the fence's storage capacity by as much as 200 percent. Through the wonders of aerodynamics, the bottom gap generates a jet of wind that keeps snow from immediately burying the fence and thereby rendering it less effective and subject to structural damage.

Plastic fence lasts about five years, or longer if you take it down for the summer to minimize livestock damage and deterioration from ultraviolet rays. If you plan to cut weeds while the fence is stored away, paint your posts a bright color so you won't run into them. Don't count on surveyor's ribbon, except as a temporary measure, since deer and other creatures like to make off with it.

Where there is a strong wind, a short fence gets buried in snow and requires constant maintenance. If you live in serious snow country, a more durable wood fence makes a better choice.

Canadian Fence

Canadian fence is the kind you are most likely to think of when someone mentions snow fence. Also called "cribbing" or "lath" fence, it typically consists of 1½-inch (38 mm) vertical slats or pickets, spaced 2⅕ inches (55 mm) apart and fastened together with rows of double-twisted wire, giving it a porosity of about 60 percent. It comes pre-assembled, in 25- or 50-foot (7.5 or 15 m) rolls, 4 or 6 feet (1.2 or 1.8 m) tall.

You will need the same framework for Canadian fence as for plastic snow fence. Even though it is sturdier and easier to handle, the 60 percent porosity of Canadian fence, combined with its vertical configuration, makes it 25 percent less efficient at trapping snow than either a plastic fence or a Wyoming fence.

Wyoming Fence

The 6-foot-tall (180 cm) horizontal board fence known as "Swedish" fence was used for snow control in the U.S. from at least 1885 until 1971. Subsequent research at Laramie led to substantial refinements and the development of what is now considered the ultimate in snow fencing — the Wyoming fence.

Built properly, a Wyoming fence is fully animal-proof, can withstand winds up to 100 mph (160 km/h), and will last twenty-five years or more. The fence may be anywhere from 6 to 14 feet (1.8 to 4 m) high,

depending on the expected depth of drifts. Fences taller than 6 feet (1.8 m), however, have special engineering requirements. Sources for further information are listed in the appendix.

You can easily build a 6-foot-tall (180 cm) Wyoming fence with 16-foot-long (5 m) pressure-treated 1x6s (2.5x15 cm) spaced horizontally 6 inches (15 cm) apart. For a long-lasting, low-maintenance fence, nail the boards to 6-inch-diameter (15 cm) 8-foot-long (240 cm) wood posts set 8 feet (240 cm) apart and 2 feet (60 cm) deep. Attach the bottom board 10 inches (25 cm) above the ground, or higher if necessary to accommodate vegetation and/or normal snow cover.

If you wish to take the fence down seasonally so you don't have to look at it all summer, construct it as a series of panels lashed to steel T-posts set 7 feet (210 cm) apart, allowing for 6-inch (15 cm) overlaps. During the summer, simply lay the panels on the ground, protected from moisture with concrete or pressure-treated wood blocks.

A Wyoming fence is as efficient as a plastic fence but is no more expensive, requires considerably less maintenance, and lasts five times longer, making it the best investment where snowdrifts are deep and the wind blows strongly.

Drift Control

Mathematical formulas have been developed to tell you exactly how to construct an efficient fence and where to put it. Although computers are used to design fences for major jobs like keeping highways clear and managing watershed resources, the fundamentals are not hard to understand.

Unlike a fence designed to keep creatures in or out, a snow fence runs in a straight line centered on the area to be kept clear. To allow enough space for a drift to accumulate, place the fence no closer than thirty times its height from the protected area.

Buildings and other solid structures exert their own influence on the flow of wind, and therefore on snow accumulation. If buildings are within the area to be protected, place the fence no closer than the sum of ten times the height of the building plus thirty times the height of the fence. Otherwise, the drift could conceivably get big enough to crush the structure.

How much snow will accumulate behind a fence depends not only on how much snow falls but also on the lay of the land, the direction of the wind, the length of the fence, its height, and its porosity. Ideal porosity, as we have already seen, is 50 percent.

How high to build your fence depends on how much snow normally is blown in your area and how important it is to trap it all. The volume of snow a fence can collect is proportional to the square of its height. A 6-foot-tall (180 cm) fence can therefore trap more than twice as much as a fence that is 4 feet (120 cm) tall.

Wind speed also plays a role. Snow is rarely lifted more than 3 feet (90 cm) off the ground, but as wind speed increases so does the height to which snow will rise. In practical terms, suppose the wind blows 26 mph (42 km/h). A 4-foot (120 cm) fence will collect only 77 percent of the blown snow under optimum conditions, while a 12-foot-high (360 cm) fence will collect 95 percent.

It is easy to determine how long to build the fence. It should be as long as the area you want to protect, with a little extra at both ends to allow for variability in wind direction. To determine exactly where to start and end your fence, run a string from each back corner of the area you want to protect to the point where it intersects the extended fence line at a 45° angle.

Minimum length is thirty times the fence's height, taking into consideration the tendency for less snow to accumulate toward the ends. This "end effect" occurs for a distance of about twelve times the height of the fence. A 4-foot (1.2 m) fence, then, doesn't efficiently trap snow for the first 48 feet (15 m) at each end. It does, however, reach 80 percent efficiency at a point five times its height from the end, or 20 feet (6 m) in.

Place the fence perpendicular to the prevailing wind, which isn't easy if the wind tends to shift around. In such a case, place the fence perpendicular to the most common wind direction. A departure in direction of up to 20° won't significantly affect snow accumulation.

The final variable — the lay of the land — is trickiest to deal with because there are a vast number of possibilities. Level terrain is, of course, ideal for constructing any kind of fence. A uniform slope of less than 10 percent will not significantly affect a

fence's efficiency. An upward slope on the windward side of the fence, or a downward slope on the leeward side, or both, increase efficiency. A significant upward hill on the leeward side works against efficiency.

Avoid placing snow fence where livestock roam. Since fences reduce wind speed, animals tend to seek shelter on the leeward side and can get buried in the accumulating drift. Although you can use snow fence to keep a penned yard clear, you will need a solid wall if you want to provide protection from wind as well as snow.

Snow fence

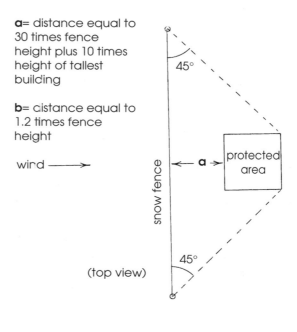

a = distance equal to 30 times fence height plus 10 times height of tallest building

b = distance equal to 1.2 times fence height

wind ⟶

snow fence

45°

protected area

45°

(top view)

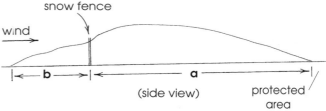

wind

snow fence

b

a

protected area

(side view)

WATER COLLECTION

If snow is a primary form of precipitation in your area, as it is in the High Plains of the western United States, set up a fence where its drift will accumulate in a stock pond. A properly placed fence can more than double the water equivalent in snow collected by a pond. Successful collection requires two conditions; frequent blizzards and a large depression in an open area.

You will need a fence that is as long as the pond is wide, placed just ahead of the pond's windward side. Ideally, dirt excavated during pond construction should be banked around the downwind side to maximize the amount of snow the pond can trap. If, as is often the case, the pond has been banked on the windward side, place the fence along the top of the embankment.

Exactly how much water you will get from trapped snow is hard to calculate because so many variables come into play. As a general rule, 10 inches (25 cm) of snow equals 1 inch (25 mm) of rain. You can expect about two-thirds of the melt to evaporate or seep into the soil, leaving a good one-third for beneficial use.

SAND OR SILT FENCE

A fence for controlling blowing sand on a desert or beach is virtually identical to a snow fence. The chief difference between them is that sand doesn't melt. Once the fence has reached its capacity, you can either cart away the trapped sand, taking care not to damage your fence, or you can erect a second fence. Since freshly accumulated sand is somewhat unstable, you will have an easier time putting the new fence somewhat in front of the old one.

Where the wind blows steadily, a 4-foot-tall (1.2 m) fence of 50 percent porosity will fill up in about a year, trapping 2 to 3 cubic yards of sand per running foot (5 to 8 cubic meters per linear meter). If wind speeds get above 36 mph (58 km/h), though, one fence won't trap much sand at all. You will need two parallel fences, the distance between them equal to four times the fences' height. Even at lower wind speeds, a double fence traps sand more effectively than a single fence.

Shifting sand, silt, and sediment in flowing water can be similarly controlled by a porous fence. The fence simply slows the water's speed, as it slows wind speed, to deposit sand or silt along the fence line.

WOVEN FENCES

▼▲▽▲▽▲▽▲▽▲▽▲▽▲▽▲▽▲▽▲▽▲▽▲▽

Woven fences are by far the sturdiest, most secure, but most expensive of all fences. They are easy to see and will control animals both large and small. Woven wire is perfect for building a secure night pen or corral, even where some other kind of fence is used for the boundary. Its chief disadvantage is that animals will lean on it, push against it, and otherwise attempt to shorten its useful life.

Woven fencing was originally made of metal wire, knotted or welded together. Today's fences are also made of plastic, woven into mesh or formed in sheets with patterned holes. Plastic fencing comes nowhere near matching the sturdiness of woven wire.

FIELD FENCE

Field fence, also called "stock" fencing, consists of horizontal wires held together by a series of permanently attached vertical wires. The spacing between horizontals depends on the purpose of the fence. In most cases, horizontals are farther apart toward the top and closer together toward the bottom.

The widest spacing between top wires is 9 inches (23 cm); the narrowest at the bottom is 1½ inches (4 cm). The narrow bottom spacing is designed both to keep in smaller stock and to give the fence extra strength, since livestock and predators alike put greatest pressure on a fence's bottom third.

Graduated spacing, compared to close spacing throughout, reduces weight and cost.

A standard roll contains 20 rods or 330 feet (100 m), but sometimes shorter rolls will contain 10 rods or 165 feet (50 m). Field fencing comes in four weights: light (14½-gauge), medium (12½-gauge), heavy (11-gauge), and extra heavy (9-gauge). The heavier the wire, the more durable the fence, and the more expensive.

For additional strength, the top and bottom horizontals are usually heavier than other line wires, called "filler" wires. The only fencing in which filler wires are not thinner is "extra heavy," which has the same gauge throughout. For all woven wire, the verticals, called "stays" or "pickets," are the same gauge as the fillers.

Heavy and extra-heavy wire are the most difficult to work with, especially if the stays are 6 inches (15 cm) apart. This wire is so heavy and stiff that you need a pickup truck or tractor to install it. Lightweight wire is the cheapest but least durable and requires extra maintenance. Most field fences are of medium weight.

Like other fence wire, field fencing may be either galvanized or aluminum coated. You can expect it to last from eight to ten years. Most brands are made of medium-hardened steel. The hardness of the steel affects the amount of springiness built into the fence. Soft steel has little intrinsic springiness and will sag no matter how carefully you stretch it.

Spring action is developed through tension curves — wavy crimps that stretch when the fence is pulled taut. Tension curves give horizontals room to stretch and shrink as the weather changes. Without them, properly stretched wire would either break or pull out its anchor posts in cold weather.

WOVEN WIRE GAUGES

	Top & Bottom	Filler
Lightweight	11	14½
Mediumweight	10	12½
Heavyweight	9	11
Extra-heavyweight	9	9

Stays

Stays are spaced either every 6 inches (15 cm) or every 12 inches (30 cm). The narrower spacing deters hogs and most dogs; the wider spacing prevents small stock like sheep and goats from getting their heads stuck. Stays are fastened with one of three kinds of joint: welded, hinge, or stiff-stay.

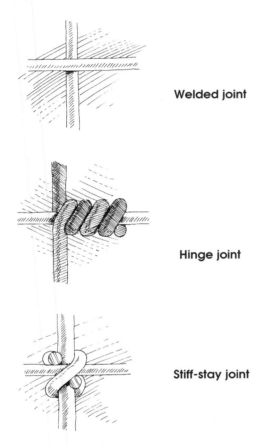

Welded joint

Hinge joint

Stiff-stay joint

Welded joints are the simplest, cheapest kind, involving straight verticals, spot-welded to the horizontals. Welded joints don't hold up well, tending to pop loose under stress or over time.

Hinge joints involve short pieces of wire individually cut and wrapped at each horizontal. The stays fold up, accordion style, when the fence gets mashed down. They are supposed to spring back into position as soon as the pressure is released. In actuality, it is extremely difficult to keep a hinge-joint fence properly stretched. Besides encouraging the fence to sag, hinge joints sometimes slip sideways, skewing the stays out of vertical.

Stiff-stay joints, also called "square knot" joints, involve one-piece vertical stays that are crimped and tied wherever they meet a horizontal. These joints won't slip. They do allow the stays to bend if the fence gets mashed down. Stays that are bent and straightened often enough eventually break. Stiff-stay joints give a fence vertical strength, reduce sag, and increase flexibility in hilly country.

Styles

Attached to each roll of field wire is a tag displaying a three- or four-digit "design" or "style" number. The last two numbers on the right represent the height in inches. The one or two numbers preceding the height specify the number of horizontal wires. An "832" fence, for example, has 8 horizontals and is 32 inches high. A "1047" fence has 10 horizontals and is 47 inches high. For a given height, the more horizontals a fence has, the stronger it is.

Some manufacturers include additional numbers such as 1047-12-11. The first number means the same as before. The second number indicates the spacing between stay wires, in this case 12 inches. The third number tells you the gauge of the filler wires, in this case 11.

The last number sometimes becomes a set of numbers separated by a slash — such as 9/11 — telling you that the top and bottom horizontals are 9-gauge and the fillers are 11-gauge. The entire design number then looks like this: 1047-12-9/11.

In countries using metric, style is designated by a series of three numbers separated by slashes, for example 8/80/15. The first number represents the number of line wires, the second gives the fence's height in centimeters, the third tells you the distance between stays in centimeters. When wire gauge is included, the numbers may all be in millimeters, like this: 13/1900/150-2.5 mm. Distributors who import woven wire usually retain the slashes when they convert from metric.

Standard woven wire ranges in height from 26 to 55 inches (66 to 140 cm). Select height based on the size of the animals you want to confine and on their ability to jump. Manufacturers tend to overstate required heights, which makes good sense from their point of view — the taller the fence, the more you pay for it.

For goats, you will likely need a 5- to 6-foot (150 to 180 cm) fence. For sheep and hogs, 26 inches (65 cm) is generally tall enough. For most horses and llamas, 45 inches (115 cm) is sufficient. For cattle, 35 inches (90 cm) is plenty, although cattle are so easy to confine that woven wire isn't cost-effective unless you keep other animals as well.

If you have different kinds of animals together, combine the tallest height you need with the closest wire spacing. Woven wire comes in such a variety of heights and spacings that you will have no trouble finding exactly what you need.

Woven wire up to about 32 inches (80 cm) is relatively easy to handle. Taller wire becomes more difficult. Many fencers add height by stringing barbed, tensioned, or electrified strands at the top. Barbed or electrified wire at the top discourages animals from leaning on the fence or trying to climb over it or crushing it down.

Run the first strand along the top 2 inches (5 cm) above the woven wire. If you need more height, make the second strand 6 inches (15 cm) above the first. A third strand should go 8 inches (20 cm) above the second.

Scare wires, as described in chapter 9, can be used to keep stock from rubbing or leaning against the fence, or from digging or burrowing under it. To prevent the latter, some people raise woven wire 3 inches (8 cm) above the ground and run a strand of barbed wire along the bottom.

Installation

Fence-building machines can install woven wire, complete from driving posts to unrolling and stretching the wire, and even splicing and attaching it. With one of these machines, a crew of three can construct 1 mile (1.6 km) of fence in a day. If you have a lot of fence to build, it may be worth your while to locate a custom fencer in your area who uses such a machine. A good place to start your search is at a farm store or builders' supply outlet.

You will need single-span anchor assemblies for your woven wire fence. Set anchor posts at least 4 feet (120 cm) deep, line posts at least 2 feet (60 cm) deep. Allow enough post above ground to accommodate the wire's height plus a couple of extra inches — more, if you will be running additional

strands above or below the woven wire.

Space line posts 14 to 20 feet (4 to 6 m) apart, depending on terrain — wider spacing for straight, level runs, closer spacing for uneven land. Place an extra post at any serious dip or rise. Use steel T-posts and clips or wood posts and staples, following the stapling procedure outlined in chapter 2.

Place woven wire on the side of the posts experiencing greatest pressure, putting the brunt of pressure against the posts rather than the fasteners. Along a curved fence, place the fencing on the *outside* of the curve. If the fence line curves back and forth, the wire will sometimes be on the inside of the posts and sometimes on the outside.

Begin by laying the roll of wire on the ground where you can unroll it along the fence line, with its bottom end — where the grid is smallest — closest to the posts. If you manage to unroll from the wrong direction, flipping the wire over won't be easy. If the fence line is curved, start at the end with the least amount of curvature.

Leaving at least 2 feet (60 cm) extending beyond the first anchor post, unroll the wire past the first three lines posts, letting it lie flat along the ground. Go back to the anchor post, lift the wire so it clears the ground by about ½ inch (1 cm), and fasten the top horizontal to the post.

Remove the first stay by cutting it between horizontals and slipping off the loosened joints. Remove a second stay, if necessary. Wrap the top horizontal around the anchor post and, using pliers or a splicing tool, wrap it around itself. Don't count on staples or clips alone to hold the wire — staples pop loose, and both staples and clips let woven wire get pulled out of shape. Fasten and wrap the bottom

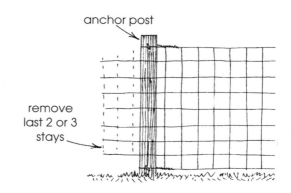

Attaching woven wire to anchor post

horizontal next, then the wires in between.

Continue unrolling the wire toward the next anchor assembly. If you have to splice two sections together, either wrap the horizontals around each other or connect them with one of the wire splicers described in chapter 5. If you use compression sleeves, slip the two ends into a sleeve until they overlap by 2 inches (5 cm) and crimp.

Wraps are the cheapest splicing method, but also the weakest. For wrap splices, cut adjoining sections of fence so the horizontals on both sides extend 6 inches (15 cm) beyond the stays. Bring the two sections together until the stays meet.

SPLICING WOVEN WIRE

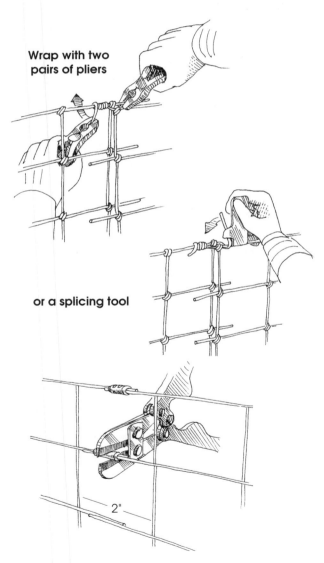

Wrap with two pairs of pliers

or a splicing tool

2"

or use compression sleeves

Using a splicing tool or two pair of pliers, wrap each end around the other section's corresponding horizontal.

Stretching

Woven wire works well only if it's stretched taut. For this you will need two wire stretchers and a dummy post, both described in chapter 5. After loosely affixing the wire to the line posts, attach a clamp bar at the point where the wire touches the last anchor post. If you don't have a bona fide clamp bar, bolt or strap two 2x4s (5x10 cm) together with the wire sandwiched between them or else weave a steel pipe or rod through the mesh.

Attach one wire stretcher to chains or cables stretched between the top of the clamp and the top of the dummy post. Attach another stretcher at the bottom. Don't try to get by with only one stretcher — you might end up warping your wire. Tighten the two stretchers very slowly, distributing the tension evenly.

With the wire partially stretched, walk along the fence and shake the wire away from the posts to make sure it isn't hung up. Then continue stretching until the tension curves are flattened by about one-third. If you go too far and flatten the curves as much as a half, the fence will lose elasticity. Overstretching can also cause horizontals to break if the temperature varies widely, since a 50°F (28°C) change can increase tension by 50 percent. When the wire is properly stretched, the verticals should remain plumb.

With the wire stretched, cut loose the middle horizontal, remove a stay or two, wrap the wire around the anchor post, and twist it around itself using pliers or a wire splicer. Next, cut and wrap the horizontal halfway between the middle wire and the top. Repeat with the horizontal halfway between the middle wire and the bottom. Working from the bottom up, secure the remaining wires. When you cut the last wire, stand back so the falling stretcher won't land on your toes.

After you have attached the wire to the second anchor post, work back up the fence line, fastening the top wire to line posts. Check to be sure the bottom clears the ground by ½ inch (1 cm). Return along the fence line and attach the bottom wire to all the line posts. On your third pass, staple or clip the wire to each line post about every 6 inches (15 cm) between the top and bottom wires.

If the fence runs through a dip, the top will be slack. If it goes over a rise, the bottom will be slack. In any major dip or rise, set an anchor assembly so you can cut and wrap the wire to adjust it to the changing angle.

If the slack is minor, you may be able to take it up with a crimping tool. You'll need a crimping tool, anyway, to restore tension when your fence starts to sag as the result of livestock pressure or the impact of ice and snow.

If there is more slack than a crimping tool can handle, cut the top (or bottom) two or three horizontals and remove the stays on both sides. Thread the ends of the horizontals through holes drilled into two 20-inch (50 cm) lengths of ½-inch-diameter (1 cm) galvanized pipe, and secure them with wraps. Connect the two pipes with two turnbuckles, as illustrated, and tighten the buckles until the fence is taut.

Don't forget to provide your woven wire fence with lightning protection, as described in chapter 5.

WOVEN HIGH-TENSILE WIRE

Woven high-tensile wire was developed in the early 1970s for deer farming in New Zealand. In the U.S. it is used primarily for exotic animal and game farming and to protect expensive horticultural projects from wild deer. It is especially suitable where animal pressure is great because of close confinement or high populations of wildlife.

A woven high-tensile fence combines the strength of high-tensile wire with the dependability of woven wire. Compared to standard woven wire, it won't sag or stretch as readily, is more resistant to rust, and is considerably lighter, allowing rolls as long as 660 feet (200 m). It is also considerably more expensive.

Heights range from 21 to 75 inches (53 to 190 cm). Taller wire — designed for deer, elk, and game farming — requires super-strong anchor assem-

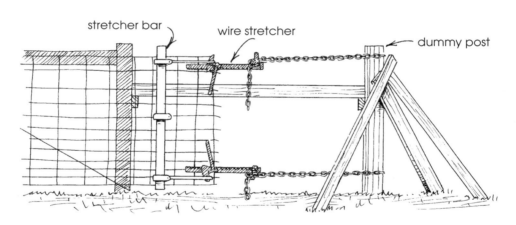

Stretching woven wire

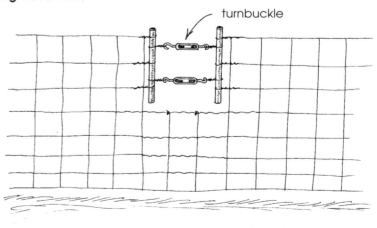

Turnbuckles take up slack

WOVEN HIGH-TENSILE FENCE GUIDELINE

	U.S.	Metric
Common livestock, llamas	10/47/9	10/120/23
Fawn; deer and game cross fence	15/61/6	15/155/15
Keep elk in; keep deer out	13/75/6	13/190/15
All-purpose perimeter fence	14/75/6	14/190/15
Fawn; deer and game perimeter fence	17/75/6	17/190/15

blies to support the extra weight. You will need double H-brace assemblies with 10-foot (3 m) brace rails. Space line posts 24 feet (7 m) apart, closer for uneven terrain. Tension the wire to about 300 pounds (1,350 N). The line wire crimps should be flattened halfway, which increases the wire's length 5 feet (1.5 m) per 330 foot (100 m) roll.

Since keeping game *in* is more difficult than keeping wildlife *out,* even the 75-inch (190 cm) wire may not be tall enough. Add height, if necessary, by running strands of high-tensile smooth wire along the top. For fallow deer, add one strand to increase the height to 84 inches (213 cm). For elk, add two strands for a total height of 90 inches (230 cm). For white-tailed deer, you may need up to five wires to raise the height to as much as 10 feet (3 m).

Game fences have been built as high as 12 feet (360 cm) by adding strands of smooth wire every 9 inches (23 cm). Even if you don't need such a tall fence, stringing high-tensile smooth wire along the top lets you get by with shorter, less expensive woven wire at the bottom.

A scare wire on the outside will deter dogs and coyotes at fawning time. A scare wire on the inside will keep stock from climbing, scratching, or pushing against the fence, letting you increase the distance between posts to 30 feet (9 m). For good visibility, keep 5 to 8 feet (150 to 240 cm) cleared of brush on the game side.

POULTRY AND GARDEN FENCE

Lightweight woven wire fencing with small openings is called "poultry and rabbit," "small stock," or "lawn and garden" fencing. It is used to fence yards, gardens, pets, poultry, and small stock and to make trellises for vegetables and flowers.

Bottom horizontals may be as close together as 1 inch (2.5 cm); verticals may be 2, 4, or 6 inches (5, 10, 15 cm) apart. Rolls range in height from 28 to 72 inches (71 to 183 cm) and contain 40, 50, or 100 feet (12, 15, 30 m).

Garden fence is also available in a welded diamond pattern designed to resemble chain link except that it is cheaper and easier to install. The wire is either galvanized or vinyl coated to offer color options.

For a light-duty fence, you can get by with steel T-posts and diagonally braced anchor assemblies. Drive posts 10 to 16 feet (3 to 4.8 m) apart and at least 18 inches (45 cm) deep. Unroll the wire with the bottom edge toward the posts. Secure the end to an anchor post. Sandwich the other end between two 1x2s (5x10 cm), the height of the fence, held fast with bolts and wing nuts.

Attach a rope to the top and bottom of this stretcher bar and have a partner pull while you work down the fence line, clipping the wire to the posts. Pull the wire just enough to keep it from sagging. If necessary, take up future sag by crimping horizontals with pliers wrapped with tape to protect the wire's coating.

If you are fencing a lawn, install the wire 1 inch (2.5 cm) above ground so you can easily cut the grass with a power trimmer. If you are fencing a garden, bring the wire as close as possible to the ground so varmints can't squeeze under. To foil munching rabbits, bury the bottom at least 1 inch (2.5 cm) below ground. To deter climbers, run scare wires along the top and bottom.

HORSE FENCE

Woven wire designed specifically for horses is very closely spaced to keep the horses from "walking down" the fence or getting a hoof caught while trying to step through. Rather than having graduated spacings, this wire is evenly spaced from top to bottom. Because of its high cost, it is used primarily for pricier equines.

While neither horses nor anything else can get through, you will need a solid rail or a scare wire along the top so they can't crush the fence down. The rail — or a strand of hot tape — also increases the fence's visibility.

So-called "square knot mesh" or "nonclimbable fence" is made of heavy-duty wire, has 2x4 inch (5x10 cm) openings, ranges in height from 36 to 72 inches (90 to 183 cm), and comes in 100- and 200-foot (30 and 60 m) rolls. The same style of wire is also made in 1x6 inch (2.5x15 cm) and 2x2 inch (5x5 cm) mesh, the latter in lighter wire gauges.

An attractive alternative is diamond mesh fence, woven in tiny triangles that form diamonds where the bases come together. Horizontals consist of two twisted smooth wires, spaced 4 inches (10 cm) apart. Additional smooth wires are woven back and forth between them to form triangles measuring either 2 or 4 inches (5 or 10 cm) at the base. Wire gauge ranges from 11 to 16, height varies from 50 to 72 inches (127 to 183 cm).

Diamond mesh has been around since the turn of the century and is the forerunner of chain link.

CHAIN LINK

Chain link, invented in Germany in 1859 and first produced in the U.S. in 1897 by the Hohulin brothers of Illinois, is the Cadillac of fences in more ways than one. It does the job better than any other fence with less maintenance, but it costs more than most people can afford. Since it will control any kind of predator or livestock except young fowl, it is the ultimate fence where absolute control is essential.

Standard mesh size is 2 inches (5 cm). Security mesh, also called "mini-mesh," comes in a variety of tighter weaves for building escape-proof pens to contain small or exotic animals. Rolls contain either 25 or 50 feet (7.5 or 15 m). Heights range upward from 36 inches (90 cm). A nonprofessional installer would be wise to avoid working with anything over 5 feet (150 cm).

Standard chain link is woven from 11-gauge steel wire and may be galvanized, aluminized, or vinyl coated in a selection of colors. It is installed on steel posts, set in concrete and evenly spaced 8 to 10 feet (2.5 to 3 m) apart, depending on whether a steel rail is used to keep the fabric taut.

Amateur fencers often omit the top rail because it requires line posts to be of exactly equal height and perfectly plumb. If you leave off the rail, cap all your line posts so they won't collect rain water.

You can install chain link yourself if you have some handyman know-how and a knack for meticulous detail — this fence has a myriad of little parts to contend with. When you buy materials, the supplier will help you determine the size, kind, and number of parts you need. Bring along a layout showing exact dimensions, locations of corners and gates, and fence height.

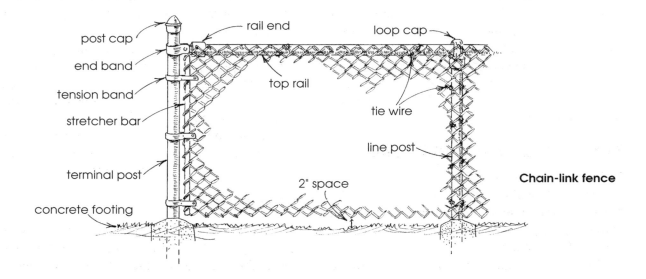

post cap
end band
tension band
stretcher bar
terminal post
concrete footing

rail end
top rail
2" space

loop cap
tie wire
line post

Chain-link fence

Some areas have stores devoted to owner-installed chain link, where you can get all the tools and moral support you need, as well as learn the lingo. In chain link parlance, for example, an end post is called a "terminal" post and chain link mesh is called "fabric."

Given the cost of materials, you may be happier with the results if you hire an experienced installer, especially if your yard is anything but level with right-angle corners. As a compromise, get a pro to install the fence and get yourself hired as a helper.

Stock Panels

Stock panels consist of welded wire so stiff it can't be bundled in rolls, so it is cut into 16-foot (4.8 m) lengths. Expensive but strong, stock panels can withstand the abuse inherent in closely confined areas and hard-to-control animals. Because stock panels are expensive, they are used primarily to create holding pens.

Horizontals and verticals are both made of ½-inch-thick (6 mm), 3-gauge galvanized steel wire. Stays are spaced 8 inches (20 cm) apart. Panel height, number of horizontals, and their spacing vary with brand and style.

Hog panels are 34 inches (86 cm) high, cattle panels 52 inches (132 cm). Other styles, variously called "ranch," "economy," and "combination" panels, are the same height as cattle panels but differ in their number of horizontals. The spacing between the horizontals ranges from 2 inches (5 cm) at the bottom to 6 inches (15 cm) at the top.

You can avoid cutting panels (not an easy task) if you make each side of the fenced area some multiple of 16 feet (4.8 m). Attach the panels to 8-foot-long (240 cm), 5-inch-diameter (13 cm) wood posts set 8 feet (240 cm) apart in concrete. Because no tensioning is involved, and the stiff panels lend stability, you won't need anchor assemblies.

After the concrete footers have set up, drive a nail into each post and temporarily hang the panels so they clear the ground by 2 inches (5 cm). Nest the panels together by turning every other one so the stays face inside, facing the stays outside on alternate panels.

Most people use barbed stock-panel staples to fasten panels to wood posts. Attach them to the *inside* of posts except at corners, where they should meet at the outside. For extra strength, lash the panels to corner posts with smooth wire.

As a substitute for staples, which tend to pop loose under pressure, you can affix each panel with three 7-inch-long (18 cm), ⅜-inch (9.5 mm) J-bolts — one at the top, one at the bottom, and one in the middle. Poke each bolt through a ⅜-inch (9.5 mm) hole drilled through the post. With the bolt positioned over a wire, thread a washer and nut on the other end and tighten enough to pull the bolt's head snug to the post.

Stock panels are best suited to level or gently sloped terrain. On a slope, set posts perpendicular to the hill and tilt the panels to follow the lay of the land.

Panels bow naturally and are therefore easy to shape into a round corral. Square or rectangular enclosures are, however, more common. To make sure the enclosure is regular in shape, set stakes at the corners and measure one diagonal, then the other. The two should be the same.

POULTRY NETTING

Poultry netting, also called "hexagonal netting," "hex net," or "hex wire," consists of thin wire, twisted and woven together into a series of hexagons, giving it a honeycomb appearance. This lightweight fencing is used primarily for backyard projects, sometimes in conjunction with a scare wire to deter raccoons.

Mesh size ranges from ½ to 2 inches (1 to 5 cm). The smaller the mesh, the stronger the fence. The smallest grid, called "aviary" netting, is made from 22-gauge wire and is used to pen quail and other small birds, to house chicks, and to prevent small wild birds from stealing poultry feed.

One-inch (2.5 cm) mesh, woven from 18-gauge wire, is commonly called "chicken wire." It is used to pen chickens, pigeons, pheasants, turkey poults, ducks, and goslings and to protect gardens. Rolls range in length from 25 to 150 feet (7.5 to 45 m), in height from 12 to 72 inches (30 to 183 cm). The shortest wire is used to reinforce the lower portion of a woven wire or rail fence to keep small animals from slipping through.

So-called "turkey netting," made of 20-gauge wire, has 2-inch (5 cm) mesh and is used for penning turkeys, peafowl, and geese and for trellising annuals. Heights range from 18 to 72 inches (45 to 183 cm), length from 25 to 150 feet (7.5 to 45 m). Mesh this large is difficult to stretch properly. For a tall fence, therefore, fencers often run two narrow rolls, one over the other. Either staple the butted edges to a rail or fasten them together with cage-making rings crimped with a tool designed for the purpose, called a "clincher" (available at farm stores and small-stock suppliers).

A less common variation, called "rabbit netting," has 1-inch (2.5 cm) mesh at the bottom and 2-inch (5 cm) mesh toward the top. It comes in 25-foot (7.5 m) rolls, is 28 inches (70 cm) high, and is used to pen chicks and poults (baby turkeys) and to keep rabbits out of gardens. Another uncommon variation has 1½-inch (4 cm) mesh, is made of 16-gauge wire, and is used for penning dogs and foxes.

In the metric world, common grid sizes are 3, 4, and 5 centimeters; common lengths are 50 and 100 meters. The 3-centimeter grid is used for tree guards, 4-centimeter for rabbit fencing, and 5-centimeter for poultry. Netting size is designated by three numbers: 90x5x1.0. The first number represents height in centimeters; the second is the grid size, also in centimeters; the third is the wire gauge in millimeters.

Unless you treat hex wire with great care, don't expect it to last more than five years. Options in protective coating are galvanizing and vinyl. Some brands are galvanized before being woven, some afterwards. The former is cheaper but should be used only under cover, since it rusts rapidly in open weather. Plastic-coated wire is a bit more rust resistant, and some people find its colors more attractive than plain metal.

Netting is relatively easy to put up but it tears readily, and slight tears grow into big holes. Netting also tends to sag. For a garden fence, clip or lash net to posts set no farther apart than 8 feet (240 cm). T-posts are most commonly used, but smooth fiberglass posts won't rub and hasten rusting.

For a pet or poultry pen, erect a stout framework of closely spaced wood posts with a baseboard and a top rail for stapling. Taller fences need a rail in the middle as well, to keep the wire taut. Hand-stretch the mesh by pulling on the tension wires — those wires woven in and out at the top and bottom of the netting. Taller netting has additional intermediate tension wires.

To avoid snagging skin and clothing, fold cut ends under before stapling them down, especially around gates. Although a solid baseboard helps keep burrowers out of a garden or henhouse, some fencers bury the bottom 6 to 12 inches (15 to 30 cm) as an extra precaution. Unless the wire is vinyl coated, coat it with roofing tar to slow rust.

APRON FENCE

Apron fencing is an option to digging a trench and burying the bottom portion of a net fence. Also called "beagle netting," apron fencing is hex wire with an apron hinged to the bottom. The apron consists of 1½-inch (4 cm) grid, 17-gauge hexagonal netting, 12 inches (30 cm) wide. It is designed to keep raccoons and foxes from burrowing into poultry yards and to keep pets from burrowing their way to freedom.

Set posts 6 to 8 feet (180 to 240 cm) apart. Cut and lift the sod along the *outside* of the fence line if your goal is to keep burrowers out, or along the *inside* if you want to keep them in. Install the fence with the apron portion spread horizontally along the ground, and replace the sod on top. The apron will get matted into the grass roots to create a barrier that discourages digging.

Use this concept to create your own apron fence with any roll of 12-inch-wide (30 cm) hex wire, clipped or lashed to the bottom of a hexagonal net fence. Whether you buy apron fencing or devise your own, the chief disadvantage is that soil moisture causes rapid rusting and the apron will have to be replaced every couple of years.

PLASTIC MESH

Plastic fencing in various forms has been around since the 1960s. Except for temporary use, such as controlling crowds around construction sites, it hasn't become popular for two reasons: it is subject to ultraviolet deterioration and it requires an expensive framework to function as an effective barrier.

The chief advantage of plastic fencing is that it is lighter than metal and therefore easier to handle. To look at but one example, a 4-foot-wide (120 cm), 350-foot (107 m) roll of plastic poultry netting weighs about 25 pounds (11 kg) compared to 60 pounds (27 kg) for a 150-foot (45 m) roll of galvanized hex wire of equal height.

Plastic fencing comes in various colors. White and orange are easiest to see; black is the most resistant to ultraviolet deterioration. Depending on the amount of ultraviolet inhibitor added, the life expectancy of plastic fence can be as little as three years or as much as ten. No modern version has been around long enough for us to know its true longevity. What *is* known is that the more inhibitor plastic fencing contains, the more it costs.

Mesh sizes and shapes vary widely, with new styles coming out and old ones disappearing faster than they can be described. Rolls range from 4 to 7 feet (120 to 210 cm) wide, and 50 to 164 feet (15 to 50 m) long. Extremely lightweight netting comes in even wider, longer rolls.

For added convenience, some brands have post sleeves that are factory-sewn. To reduce sagging, some have a tension load line running through a top hem. At each wood post, you pull the line through a hole in the hem, twist it into a loop around the outside of the post, and slip the loop into a ½-inch (6 mm) groove cut into the post top.

Plastic fencing may be woven from twine or it may consist of a solid sheet with patterned holes. The latter is sometimes oriented, or stretched, in two directions for greater strength. Mesh that has been oriented in only one direction is cheaper, but not as strong. Mesh that hasn't been oriented at all is cheapest and weakest.

Uses

Uses for plastic fencing range from housing baby birds to retaining full-grown horses. Plastic fences for controlling snow, sand, or silt are discussed in chapter 10. Given the right framework, plastic can be substituted for any woven wire of comparable weave. It can also be attached to a rail fence to keep small creatures from slipping through.

Since plastic doesn't rust, the bottom can easily be buried to control burrowing. Plastic won't, however, deter persistent chewers, including hungry dogs, foxes, or weasels, unless augmented with a scare wire.

Plastic material designed for the purpose doubles as a sunscreen to reduce heat by as much as

15° F (8° C) or as a windbreak to reduce wind velocity by half. Without a substantial framework, plastic fencing tears easily in high winds.

To give you some idea what "substantial" means: one Midwestern stockman protects his cattle from winter winds with plastic fence sandwiched between lath and recycled telephone poles set 8 feet (240 cm) apart in concrete.

Procedure

Posts for plastic fencing can be wood, fiberglass, or smooth metal rod. Set line posts 10 to 12 feet (300 to 360 cm) apart; 6 to 8 feet (180 to 240 cm) in windy areas. Diagonal anchor assemblies are sufficient for most situations. Single H-brace assemblies hold better in high winds. Guys make the handiest bracing for a temporary fence.

Fasteners for rod posts can be nylon twine, or you can weave the posts through the mesh and drive them into the ground. For a windy area, use wood posts and staple, nail, or lash the mesh between the posts and pieces of lath, taking care not to staple or nail into the mesh itself.

For a permanent fence, increase rigidity with a bottom rail of fiberglass, smooth steel, or wood. A top rail would also improve rigidity. If you use rods, weave them through the mesh and lash them securely to the line posts. For a sturdier framework, nail wood rails to wood posts.

Avoid abrasive posts, rails, or ties that can cause the fencing to tear loose in a high wind. If you use steel T-posts, which characteristically have rough edges, sandwich mesh between two pieces of lath and use smooth wire to lash the lath to the posts.

Some manufacturers provide splice rods for securing two ends together (overlap the ends and weave the rod through) or for attaching fencing to an end post (wrap the fence around the post, then back on itself, and weave the rod through). As an alternative to splice rods, use thin fiberglass rods or sandwich overlapping ends between two pieces of wood, lashed together. Lightweight netting can simply be stitched together with plastic or nylon twine.

To build a plastic fence, wrap one end around an end post and secure it. Unroll the mesh along the fence line, weave a rigid rod through the other end, and stretch. If you put up the fence in summer, allow slight slack to accommodate cold weather contraction. In winter, stretch as taut as possible, since the plastic will expand come summer. Wrap the fencing around the far end post and secure it.

Trellising

Plastic mesh with grids in the 6-inch (15 cm) range can be used to trellis annuals such as beans, peas, melons, and squash and to keep ripening tomatoes off the ground. Set two sturdy support posts 25 feet (7.5 m) apart and stretch a length of plastic or plastic-coated line wire between their tops. With nylon string, lash one end of the net to one post. Then lash the net along the wire so it hangs like a curtain from a rod. Pull the mesh taut without stretching it out of shape and lash the far end to the second post.

The biggest problem with plastic trellising is disposing of it at season's end. You can't save it for re-use by burning off dried vines with a torch, as you can with metal wire. Your options are to burn the whole thing (adding toxins to the atmosphere), toss it in the trash (adding to our growing landfill problem), or look into the possibility of recycling. If you favor a reduction in the use of petroleum products, avoid plastic fencing altogether.

PLASTIC FENCE

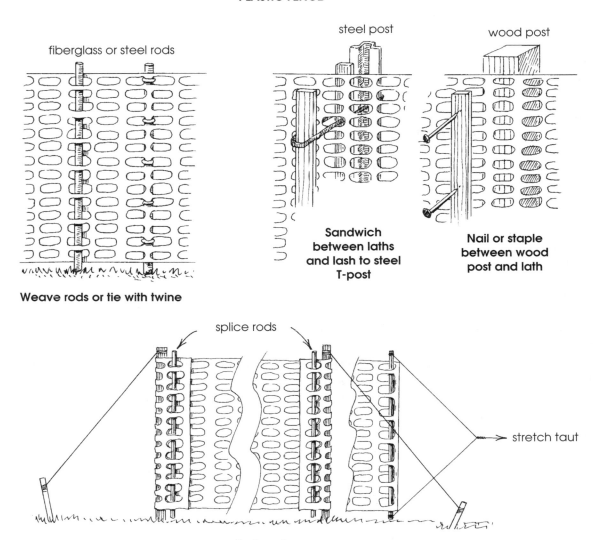

fiberglass or steel rods

Weave rods or tie with twine

steel post

Sandwich between laths and lash to steel T-post

wood post

Nail or staple between wood post and lath

splice rods

stretch taut

End posts

RAIL FENCES

▽▲▽▲▽▲▽▲▽▲▽▲▽▲▽▲▽▲▽▲▽▲▽▲▽▲▽▲▽▲

Tradition and rustic charm are the main reasons rail fences remain popular in areas such as Kentucky's horse country. These fences look sturdy and are plainly visible to horses and humans, giving them value as physical barriers. They work best on level or gently sloped terrain but are difficult to build in hilly country, where they neither look right nor hold up well.

A rail fence made of wood, steel, concrete, or vinyl will control only the easiest-to-confine animals — horses, mules, llamas, cattle, and sheep — unless you are willing to augment the rails with woven wire, plastic mesh, or electrified scare wires. Alternatively, you might opt for an electrified high-tensile rail fence.

WOOD RAIL FENCES

In the history of fencing, rail fences come first, built by early settlers who had a surplus of timber after clearing wooded land. Wood fences aren't as common today as they once were, partly because of their cost and partly because durable hardwoods are rapidly disappearing. A fence made of fast-growing hybrid softwood isn't nearly the same good investment, considering the work and expense that goes into construction.

The best wood fence warps, splinters, breaks, rots, and gets chewed up. It requires hours of maintenance, especially if it is painted or built in a humid area where posts and rails need frequent replacement. This is not the most durable kind of fence and can be the most expensive, unless you harvest your own wood.

How long a wood fence will last depends primarily on how much heartwood it contains. You can't control the amount of heartwood in rails unless you harvest them yourself. Commercially prepared rails contain far more sapwood than they used to, causing them to decay more rapidly.

Good heartwood rails of black locust, cedar, chestnut, redwood, or white oak last as long as 100 years. Heartwood rails of ash, hickory, maple, spruce, tamarack, or white pine last from 25 to 50 years.

Because good heartwood is both scarce and expensive, pressure-treated lumber is offered as an alternative. Pressure treatment increases the durability of a rail fence by twenty to thirty years. Without pressure treatment, some wood fences last five years or even less. Fir, hemlock, and pine are the most common pressure-treated species.

Posts and any rails that touch the ground should be treated for "ground contact." An "aboveground" rating is sufficient for suspended rails. Use corrosion-resistant fasteners, as described in chapter 2, to avoid deterioration through interaction with salts in the preservative.

Many hardware stores and lumberyards carry components for rail fences, usually made of pressure-treated yellow pine or mixed hardwoods. Posts are cut to size and rails are uniform in length. The style code for a three-rail fence with 7-foot 6-inch posts, for example, looks like this: 3R76.

Virginia Zigzag

Early fences were made of slender logs, larger ones split so they'd go farther. The rustic appearance of this kind of fence makes it picturesque in the right setting, but the rough, splintered rails are definitely not user-friendly. You can buy split rails from a lumberyard or building supplier, but they will be sawn and won't look as authentic as hand-split rails.

The easiest and the most primitive rail fence to construct is the Virginia zigzag, also called a "worm" fence. It is the easiest to knock, too, over unless you add posts for support. Rails consist of round or split poles set at 90° angles and stacked as high as 6 feet (180 cm).

DURABILITY OF WOOD RAILS

Wood Species	Decay Resistance	Warp Resistance	Bending Strength
Ash, black	poor	poor	fair
Aspen	poor	—	—
Baldcypress, old gr.	good	good	good
Baldcypress, young gr.	fair	fair	fair
Basswood	poor	poor	poor
Beech	poor	poor	good
Birch	poor	poor	good
Butternut	poor	poor	poor
Catalpa	good	good	good
Cedar, red	fair	fair	poor
Cedar, white	fair	fair	poor
Cherry, black	good	good	fair
Chestnut	good	fair	poor
Cottonwood	poor	poor	poor
Cypress	good	good	good
Elm	poor	fair	good
Fir, Douglas	good	fair	good
Hackberry	poor	—	—
Hemlock, eastern	fair	fair	fair
Hickory	poor	fair	fair
Larch	fair	fair	good
Locust, black	very good	good	good
Locust, honey	fair	poor	poor
Maple	poor	poor	good
Mulberry, red	very good	good	good
Oak, black	poor	good	good
Oak, red	good	poor	good
Oak, white	good	good	good
Osage orange	very good	good	good
Pine	poor	fair	poor
Pine, southern yellow	fair	good	good
Poplar, yellow	poor	fair	poor
Redwood	good	good	good
Sassafras	good	good	good
Spruce	fair	fair	fair
Sweetgum	poor	poor	good
Sycamore	poor	fair	fair
Tamarack	good	—	—
Tupelo, black	—	poor	good
Walnut, black	good	good	fair
Willow	poor	poor	poor
Yew	very good	good	good

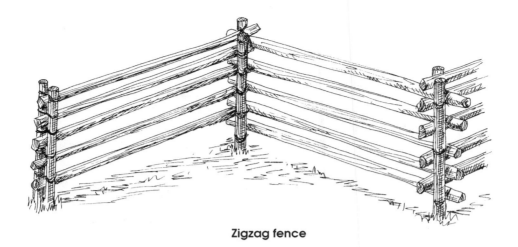

Zigzag fence

Among rail fences, the zigzag conforms most readily to hilly terrain, but it takes up lots of space and isn't considered ecologically sound because it is highly wasteful of timber. The fence remains of interest primarily for its historic charm and is often used for displaying classic livestock breeds.

This style of fence evolved in the wooded areas of Kentucky, Tennessee, and Virginia, where you can still find examples one hundred years old.

Kentucky Rail Fence

A Kentucky rail fence, also called a Virginia rail fence, is essentially a zigzag fence straightened out to eliminate the zigs and zags. It uses less timber than a zigzag fence and takes up less space, but isn't nearly as interesting to look at and is more difficult to construct on a slope.

Although traditionally built with split rails, this fence can also be made with round rails. The 11-foot-long (330 cm) rails are stacked between pairs of posts so they overlap by 6 inches (15 cm) at each end, with the bottom rails resting on flat stones. Posts, set 10 feet (300 cm) apart, are lashed firmly together in a figure eight, as illustrated.

Buck Fence

The buck fence, or "jackleg" fence, originated in the Rocky Mountains. It is a sturdy rail fence that is easy to put up and can withstand snow loads, wind, and back-rubbing animals. It is about 4½ feet (135 cm) high and very rustic.

The name comes from the crossed poles — called "bucks" or "jacklegs" — that hold the rails in place. The bucks rest on top of the ground, so no

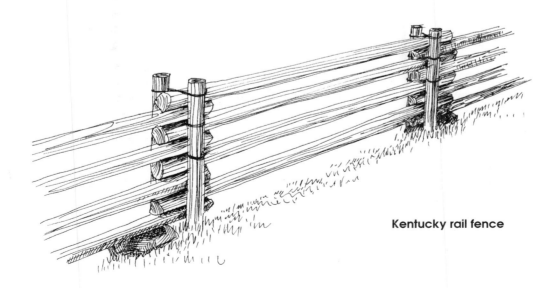

Kentucky rail fence

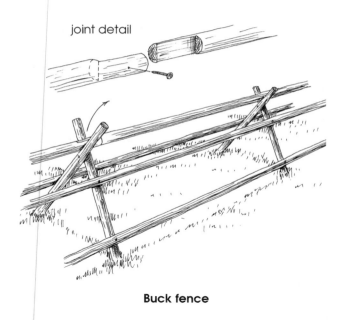

joint detail

Buck fence

post holes are needed, making this fence suitable for areas where the soil is frozen or rocky.

You will need two sets of poles, one for the bucks and one for the rails. Poles for the bucks are 6 to 8 inches (15 to 20 cm) in diameter and 6 feet (180 cm) long. You will need two per buck. Use an ax to notch the poles 18 inches (45 cm) from their tops so they fit snugly together when crossed at a 45° angle. Fasten the poles together with a 6-inch (15 cm) galvanized nail.

The poles used as rails are 4 to 6 inches (10 to 15 cm) in diameter and 11 feet (330 cm) long. For every 10 feet (300 cm) of fence, you will need one buck and four rails. Select one of the four as the top rail and cut an 8-inch (20 cm) slice from each end, as illustrated. Oriented with the sliced portions alternately upward or downward, the rails fit snugly together.

Lay the bucks along the fence line, 10 feet (3 m) apart, and lay four rails between each pair. Working with a partner, preferably two, set up the first two bucks. Place a top rail in the cradles and nail it down. Continue setting up bucks and laying in top rails, butting the sliced-away ends for stability.

With all the bucks and top rails in place, go back and nail on the lower rails. For a good fit, notch the rails and bucks where they intersect. Attach one rail 8 inches (20 cm) below the top rail. Attach another 24 inches (60 cm) below that, on the same leg. Attach the final rail on the opposite leg, 20

inches (50 cm) from the top rail.

Lodgepole pine is the wood of preference for a buck fence. It is straight, the right size, and easy to peel. You can cut your own as part of woodlot thinning, or buy it as forestry salvage. To control wind and blowing snow, lay sawmill discard slabs vertically against the rails, alternately on the inside and outside of the fence so their tops cross, and lash them down.

Post-and-Rail

Next on the evolutionary scale is the post-and-rail fence. Its appearance and strength derive from absolutely accurate post positioning. Unlike other kinds of fence where you set all the posts and then attach fencing material, here you add the rails as you install each post. Because posts rot more quickly than rails (decay occurs primarily underground), they should be pressure treated or of a more durable wood than the rails.

Set posts at least 2½ feet (75 cm) deep in concrete or backfill, as the soil conditions require. After setting the first post, place the second and third posts in their holes, but don't backfill until you have attached the rails. If you are setting posts in

POST-AND-RAIL FENCE

top view

Taper rails to fit snugly

line post   end post   corner post

Mortise profiles

concrete, avoid knocking them out of plumb by allowing the concrete to set for at least twenty-four hours. Constructing a fence one section at a time is decidedly slow work, but the resulting accuracy is well worth the trouble.

The most rustic fences have split rails with their rounded edges oriented upward to shed rainwater. Place the heaviest, straightest rails toward the top. Sturdier fences have round rails consisting of either peeled poles or milled round rails. Rails can also be made of milled lumber, either 2x4s or 2x6s (5x10 or 5x15 cm).

The number of rails you need depends on the purpose of the fence. Two rails will keep in sheep or llamas; three or four will deter horses, mules, or cattle; five will isolate stallions or bulls.

The smaller the animals, the closer the bottom rail must be to the ground and the closer together rails should be spaced. Typical spacings between rails range from 8 to 12 inches (20 to 30 cm). Typical fence heights range from 3 to 6 feet (90 to 180 cm). The top rail generally falls 3 to 6 inches (8 to 15 cm) below the post tops.

The rails run parallel to the ground and are either stacked between paired posts (as for a Kentucky rail fence) or slipped into slots cut through the posts. The latter is called "mortise and tenon" — the slot is the mortise and the rail is the tenon.

Since the rails overlap 12 inches (30 cm) at each end, they must be 10 feet (300 cm) long for posts set 8 feet (240 cm) apart. Split or round rails generally overlap one on top of the other; milled boards rest side by side. Taper the last 10 inches (25 cm) of each split or round rail for tight-fitting overlaps.

Mortising is best done before the posts are set. Mark rail positions, measuring from the top of the post down. Outline each mortise — sized to accommodate your rails — by drilling a series of holes with a 1-inch (2.5 cm) flat boring bit. Square the hole with a wood chisel and mallet and smooth it with a chisel or rasp. When you set the posts in the ground, be sure they are all the same height above the ground and the mortises are properly oriented.

Line posts must be at least three times the diameter of the rails to retain their strength despite all the holes. Corner posts should be even stouter to allow for large hollowed-out areas where rails meet at right angles. Don't mortise all the way through a corner or end post where rails end. Since line, end, and corner posts are mortised differently (as illustrated), lay out your fence on paper to figure out how many of each kind you will need.

If you like the look but not the work, buy prefab mortise-and-tenon fence components. Although the posts will already be mortised and the rails tapered, the rails may need a bit of additional trimming for a good fit.

English Hurdle

A variation on the post-and-rail fence, the English hurdle may be thought of as an early form of stock panel. The hardwood panels originated as a quick way to set up portable corrals that could be moved periodically as forage was grazed down.

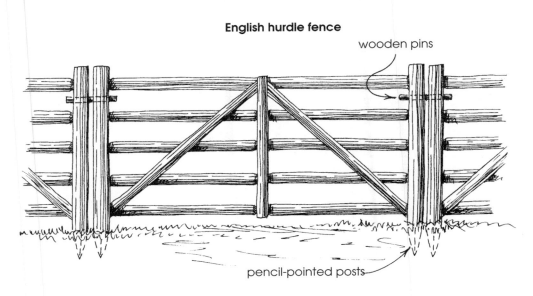

English hurdle fence

wooden pins

pencil-pointed posts

Each panel is 4 feet (120 cm) high and 8 feet (240 cm) long, with pencil-pointed posts at each end. Four or five rails are supported by one vertical and two diagonal braces. The rails, tapered at both ends, rest in mortises that go only partway through the posts.

To put up an English hurdle fence, pound the posts into the ground and tie the tops of adjoining posts together with smooth wire. For a historic look, secure the posts instead with wooden pins inserted horizontally into holes bored through them. Such pins come with ready-made hurdle kits. For convenience, substitute carriage bolts with nuts and washers.

Originally designed for confining sheep, an English hurdle fence will also hold other kinds of stock. But because the posts aren't set very deep, the fence won't withstand a lot of leaning or back scratching. Furthermore, this fence is no longer considered as portable as it was in its glory days.

Each hardwood panel weighs 65 to 70 pounds (140 to 155 kg).

Post-and-Plank

A post-and-plank fence, also called "post-and-board," is easier to put up than a post-and-rail fence and looks more formal, especially when painted. The rails consist of milled planks fastened to round or square posts.

The best planks have good bending strength, hold nails well, are warp free, and are relatively resistant to decay and weathering. Good choices include cypress, Douglas fir, redwood, Southern yellow pine, western larch, and white oak. Any of these last longer if pressure treated.

Milled boards are graded as number 1, 2, or 3, based on the number and size of the knots and other imperfections they contain. You can usually, although not always, find a stamp on each board

Post-and-plank fence on level ground

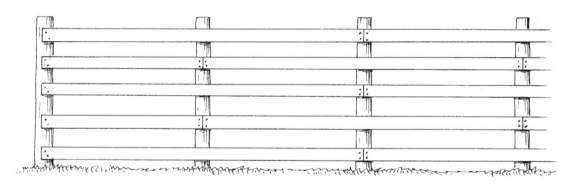

Stepped post-and-plank fence with scare wire closing the bottom gap

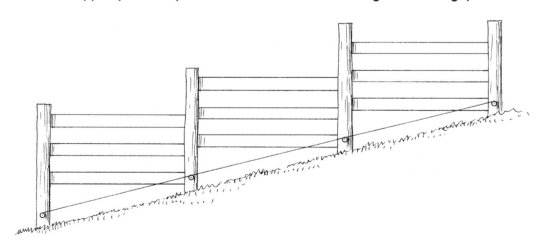

showing what grade and species it is. For the strongest, most durable rails, use nothing less than number 2 grade softwood or number 2 common hardwood (number 1 being the highest and the most expensive grade). Unfortunately, grading is often based on the best side, but in a fence you see both sides.

Since grading standards aren't uniformly followed, go to the lumberyard and pick out the boards yourself. If you have them delivered, you may get boards of lesser quality. Knots and other imperfections are more than an aesthetic problem — they collect moisture that encourages decay.

When it comes to a plank fence, you can't be both fussy and cheap. If you opt for cheap, look for a mill that sells rough-sawn lumber. The boards will be rough and splintery, as well as irregular in width and thickness, and will soak up prodigious amounts of paint. Instead of painting these rails, use stain or whitewash, or let them weather naturally.

Rough-sawn or not, boards should be fully dry before fence construction begins. Green rails will check, shrink, warp, and crack as they dry.

Construction

The posts of a typical board fence are 8 to 10 feet (240 to 300 cm) long, set 2½ to 4 feet (75 to 120 cm) deep and 8 feet (240 cm) apart. You will have to cheat and space the first line post a few inches closer to an end or corner, since rails are fastened to the *center* of a line post but to the *outside* edge of an end or corner post. On rolling terrain, place a post on each rise or in each dip. The more uneven the terrain, the closer together posts must go to make their spacing come out even.

Planks are usually 16 feet (480 cm) long, 4 to 6 inches (10 to 15 cm) wide, and 1 to 2 inches (2.5 to 5 cm) thick. The thicker the rails, the stronger the fence. In reality, 1-inch-thick rails are only ¾ inch (19 mm) thick; they soon warp and split, especially when they aren't painted.

If you are building a corral or other place of close confinement, space posts 5 or 6 feet (150 to 180 cm) apart. Make the rails 10 to 12 feet (300 to 360 cm) long, 2 inches (10 cm) or more thick, and 6 to 8 inches (15 to 20 cm) wide.

Each rail will be supported by three posts, the joints between them staggered so all joints won't fall on the same posts. When you start construction, make the topmost rail full length, cut the second rail down in half, make the third rail full length, and so on. Use full rails for all subsequent sections until you come to a corner, then fill in the ends with half-length rails.

Two to four rails are typical, attached to the livestock side. For horses — the most common animal for which board fences are built — place the top rail higher than the tallest animal's shoulder, usually 4½ to 5 feet (135 to 150 cm) but at least 6 feet (180 cm) for stallions. Make the lowest rail 14 inches (35 cm) high and space intermediate rails at least 14 inches (35 cm) apart so the horses can't get their hooves caught.

Turn the bow in crooked boards upward so your fence won't take on a saggy look. Attach the rails in one of four ways: nail or bolt them to the posts, fasten them with brackets, or use mortise-and-tenon joints, as for a post-and-rail fence.

Mortising is the least common method because it is the most work, but the fence it creates is equally strong on both sides. Furthermore, because the rails aren't permanently affixed, damaged ones can easily be replaced. Remove the damaged rail, slide the remaining rails down until you come to the end, mortise through the end post, and slide in a new rail.

Nailing is the fastest, most common method, but not necessarily the most desirable, since back-rubbing animals and changing weather eventually work nails loose. Scare wires (described in chapter 9) will help keep stock from rubbing against the fence.

Use ring-shank or screw-shank nails, as described in chapter 2, allowing two or three per end, depending on rail width. For spacing accuracy, unless you are working with a partner who can hold rails while you hammer, clamp each rail to its posts before nailing it.

Avoid splitting the grain by staggering nails instead of putting one above the other. Use posts with a flat nailing surface such as split or faced round posts or pressure-treated square posts — 4x4s (10x10 cm) for line posts, 6x6s (15x15 cm) for anchor posts.

Bolting is more expensive and more time-consuming than nailing because you have to drill a pilot hole for each bolt. However, bolts won't work loose

— they will hold as long as the wood lasts, no matter how much rubbing goes on. And bolts simplify removal and replacement of damaged rails.

Brackets, like bolts, minimize the problem of loose rails. They come in different sizes to accommodate different rail widths, and in different styles to fit round or square wood posts, galvanized pipe, or steel T-posts. Using the latter, you can put up a post-and-plank fence quickly, since T-posts are driven rather than hand set.

On level terrain, run rails parallel to the ground. For a secure nighttime corral, add a scare wire along the outside. If you want a combination fence to hold a variety of stock and keep out predators, string electrified wires between the rails.

On a slope, either run rails parallel or step them, as illustrated. If the slope is quite steep, stepping leaves a big gap on the uphill side of each post. To keep small animals from getting through, run woven wire or a scare wire along the bottom of the fence.

Finishing Touches

Instead of making all the rails parallel to each other, a popular decorative option is to cross some of the rails between the top and bottom ones to create a series of diamond patterns.

Another final touch consists of nailing a fascia board along the post tops, wide face parallel to the ground, after cutting the post tops to slope slightly toward the side on which the rails are fastened. The sloping ledge thus formed strengthens the fence. Fascia proponents claim it also protects post tops from moisture. In actuality, when two boards meet at the top of a post, they collect moisture that *hastens* post decay.

In contrast to post-and-rail fences, which are often left to weather naturally so they will blend into country landscape, post-and-plank fences have a formal look that cries out for paint. White is the traditional color for a horse fence, since it is highly visible. Black is becoming the color of choice because it is more weather resistant — a black fence requires repainting every four or five years, compared to every second year for a white fence. Other common colors include dark brown, medium green, and barn red.

ALTERNATIVES TO WOOD

Over the years, various alternatives have been devised to overcome the disadvantages of an all-wood rail fence. These materials are invariably more expensive than wood, but they are more durable.

Stone-and-Board

Posts are the most vulnerable part of a wood fence, since decay occurs primarily below ground. Stone posts (available at some building centers and large nurseries) will give you not only a more durable fence, but also one with a decidedly elegant look. On the other hand, stone posts may not be practical in rocky soil where a uniform depth is difficult to achieve.

Concrete Rail

Concrete posts offer a less expensive alternative to stone. For a super-durable fence, use concrete rails as well. A concrete rail fence won't warp or splinter, and requires a fraction of the upkeep of a wood fence.

In many areas you can purchase concrete fences ready-made and installed, should you so choose. Colors can be custom blended to match surrounding buildings or existing fences. Some brands, through vertical casting, have rails with an attractive woodlike finish on both sides.

Posts should be set 2 feet (60 cm) deep in concrete footers. Prefabricated posts have an H-shaped cross section. The rails, with spacers between them, slip one by one into the channels thus formed. Use wooden templates as measuring devices to make sure posts are perfectly plumb and the distance between them is exactly right for the length of your rails.

A concrete fence is easy to install, but the weights involved offer a unique challenge. You'll have an easier time of it if you can arrange to use a forklift or a tractor with a hydraulic hoist.

Galvanized Steel

Another durable option to wood is galvanized steel, fashioned into pipe or "boards." Like concrete, steel can't be chewed, won't splinter, and needs no paint. Use fittings to attach pipe rails to pipe posts set in concrete or opt for prefabricated panels made from welded steel tubing. As practical as they are, pipe fences aren't popular because of their stark, institutional look. Hollow steel posts and boards, on the other hand, are textured and colored to look like wood.

Vinyl

Another alternative to wood fences, popular in Europe and Canada and introduced into the U.S. in the late 1970s, is fencing made of polyvinyl chloride, often abbreviated as "PVC" or "vinyl." The number of rails, distance between them, total height, and post length vary with style and manufacturer.

Vinyl is durable and virtually maintenance free, never needs painting, holds up better than wood in humid conditions, and is impervious to weather, corrosion, insects, and decay. It is flexible enough to absorb collisions yet tough enough to resist cribbing, which leads to a damaged fence and a mouthful of splinters.

Vinyl fences are prefabricated with easy-to-assemble components and ready-made gates. Depending on brand, the rails either attach to brackets on the faces of posts or slip into holes in the sides (similar to mortise and tenon joints). To shorten a rail — for example to round a curve or go over a hill — use a fine-tooth cross-cut or saber saw.

The chief problem with vinyl fencing is deterioration caused by ultraviolet light. Resistance is influenced by ultraviolet inhibitors in the vinyl and by color. Vinyl fencing comes in traditional white, ivory, gray, brown, and black, the latter being the most resistant to ultraviolet rays.

Vinyl-Clad

Another plastic option is vinyl-clad wood fencing, which also comes in a variety of colors. Unlike plain vinyl, which is hollow at the core, this fence is made of PVC-coated kiln-dried lumber. The ends of all posts and rails are capped to seal out moisture.

This fence has both the strength of wood and the maintenance-free advantages of vinyl. Vinyl-clad lumber comes in standard dimensions and can be nailed and sawed like wood. To prevent staining, assemble a vinyl-clad fence with galvanized ring-shank or other rust-free nails.

Flexible Rail

Flexible rail fencing was designed to reduce fence-related injury to horses, the cause of more than half of all on-farm injuries. Flexible rails won't splinter and break like wood, cut or tear like wire, cause abrasions and broken bones like rigid concrete and pipe, or pop out of brackets on impact like rigid vinyl rails.

Each rail consists of a vinyl or polymer (polyethylene) strip with sleeves at the top and bottom (sometimes in the center) through which high-tensile wire is threaded. The wire may move within the sleeve or be bonded to it, depending on brand.

Free-moving wire can be tensioned separately from the sleeve. When the two are bonded, they must be tensioned together, creating stress to the sleeve, since plastic and metal expand and contract at different rates.

When wires and plastic are molded together, the rails are slipped through brackets attached to the livestock side of wood posts, like a belt threaded through loops in a pair of jeans. When the wires and plastic are independent, the wires are stapled to wood posts through the plastic sleeve.

You can electrify a vinyl-clad or flexible rail fence in the same way as you would a wood fence by installing insulators on the posts between rails and stringing hot line wires.

All-Electric

An all-electric rail fence offers a relatively inexpensive alternative to other rail options, and one that controls a wider range of livestock and predators. Developed by Common Sense Fence and consisting of fiberglass posts and high-tensile wire sheathed in round rails, it combines high visibility with all the advantages (and disadvantages) of electric fencing.

Rails consist of galvanized, painted sheet steel tubes, slotted so they can easily be dropped over the line wires. The electric pulse passes right through the metal rails. To save money, erect the fence with only a top rail for visibility, leaving the lower line wires bare or adding more rails later, as you wish.

CHAPTER 13 ▼▲▼▲▼▲▼▲▼▲▼

GATES

▼▲▼▲▼▲▼▲▼▲▼▲▼▲▼▲▼▲▼▲▼▲▼▲

A gate gets more wear than the rest of the fence, so it should be strong, well hung, well braced, and made of top-quality materials. There is no sense in investing a lot of time and money in a fence, only to weaken the whole system with an inappropriate or poorly built gate.

GATE SIZE

As important as proper gate placement (discussed in chapter 1) is proper size. A gate designed strictly for foot traffic should be wide enough to admit your wheelbarrow, garden cart, or riding lawnmower. In general, 4 feet (120 cm) is the minimum width for foot traffic.

For larger equipment or livestock, a 10- to 12-foot (3 to 4 m) gate is more appropriate. For vehicles and machinery, 14 feet (4.3 m) should be wide enough, although a 16-foot (5 m) gate may be necessary for major farm machinery, especially if the driver has to turn at the entry.

If you have any doubts about what size gate you need, play it safe and go to the next larger size. The gate's height should, of course, match your fence.

GATE POSTS

Gate posts should be sturdy enough to hold the gate's weight without leaning and permit the gate to swing freely. The wider and heavier your gate is, the more likely it will pull its post out of plumb unless the post is particularly stout, deep set, and well braced.

Gate posts should be at least 1 foot (30 cm)

longer than line posts so they can be set 1 foot (30 cm) deeper. Depending on the size and weight of your gate, the post on the hinge side — which bears most of the weight — should consist of a 6x6 (15x15 cm) or 8x8 (20x20 cm), or the equivalent, set at least 3 feet (90 cm) deep.

Support the posts on both sides with single-brace assemblies and set the hinge post in concrete. If you are hanging a double gate (described below), both posts will be hinge posts and should be set in concrete. Let the concrete cure at least seven days before hanging the gate.

To keep down weeds, minimizing maintenance and hiding places for snakes, fashion wide concrete aprons around each post. Alternatively, use vinyl landscape trimming around the entryway. Discourage digging predators and livestock by burying a concrete footer or pressure-treated post across the gateway.

FRAMED GATES

A good gate is strong enough to support its own weight. The strongest gates are framed gates made of wood or metal. They are expensive but dependable, making them the best choice for a perimeter fence or a corral, where the gate is subject to frequent bumping by animals.

Framed gates often have wide spaces through which dogs and other small animals can slip. You can eliminate the problem by lashing or stapling metal or plastic mesh fencing to the gate, or by affixing scare wires to one or both sides.

If you are fitting a new gate to existing gate posts, determine its width by measuring between the posts, inside face to inside face. Measure both top and bottom in case the posts are not plumb, and subtract an allowance for hinges and latches. Typical gate hardware takes up 3¾ inches (95 mm) for a single gate, 5½ inches (139 mm) for a double gate.

Wood Gates

The chief problem with wood gates is that they are heavy and tend to sag if they are not properly made. For the strongest, most durable gate, use pressure-treated lumber. Many states distribute construction plans for wood gates through county extension agents or state land-grant universities.

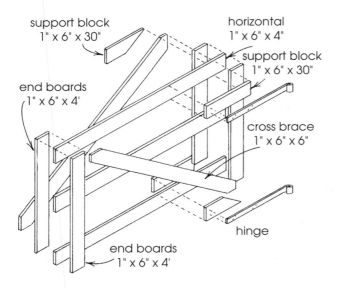

support block
1" x 6" x 30"

horizontal
1" x 6" x 4"

support block
1" x 6" x 30"

end boards
1" x 6" x 4'

cross brace
1" x 6" x 6"

hinge

end boards
1" x 6" x 4'

Framed 4-foot-wide walk-through gate

As a shortcut for a lightweight garden gate, use 1x4s (2.5x10 cm) to construct a pair of matching frames that fit your gateway. Sandwich woven wire or plastic mesh between the frames and fasten them together with wood screws. For a shortcut drive-through gate, cut a set of rails to fit your gate opening and fasten them together with prefabricated braces, called "gate irons," available at many farm supply centers.

For a more homey look, make a sturdy gate of 1x6s (2.5x15 cm). To make a 4-foot-wide (120 cm) walk-through gate, you will need three 4-foot (120 cm) rails, four 4-foot (120 cm) end boards, two 6-foot (180 cm) cross braces, and four 30-inch (75 cm) hinge support blocks.

To make a 12-foot (360 cm) drive-through gate, use 2x6s (5x15 cm). Triple the length of the three rails and increase the brace length to 7 feet (210 cm). Add a second pair of cross braces and end boards so the finished fence has the appearance of a double-X.

Assemble the gate on a garage floor or other level surface. Begin by spacing the rails evenly and sandwiching them between the end boards, as illustrated. Then position the cross braces. Place support blocks in the top and bottom corners on the hinge side, front and back, beveling two of the blocks so they will fit snugly against the brace.

When you are sure the gate is square (the diagonals should be equal), glue it with exterior glue and hold it together with clamps. Before the glue dries, drill countersunk pilot holes and fasten the gate together with flat-head wood screws. Where vehicle traffic is infrequent, consider a double gate — two narrow gates, one hinged to each gate post so their latch sides meet in the middle. Anchor one side so it can be swung open easily when you need a wider passage.

To anchor down one side of a double gate, or to keep a gate from blowing shut while you're driving through, drop a short length of steel pipe or a rod into a concrete sleeve and attach the gate to the rod. To make the sleeve, dig an 8-inch-diameter (20 cm) hole and set a pipe in the center, stuffing a bit of paper into the top end so you won't spill concrete into the pipe. Fill the hole around the pipe with concrete and smooth it flush with the ground. After the concrete sets, drop a pipe or rod of smaller diameter into the hole. Attach the gate to the pipe.

Framed wooden drive-through gate

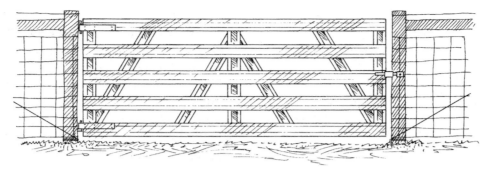

Framed metal gate

If you want to anchor one side of a double gate, pour the sleeve at the end of the gate where it meets the center of the gateway. To hold a gate open, pour the sleeve by the side of the driveway where the end of the opened gate comes to rest. To move the gate, simply lift the rod out of its sleeve.

Metal Gates

Prefabricated metal gates are available at any good farm supply store. They come in various widths and are made of either aluminum or steel. Steel makes the strongest, most durable of all gates. If you're skilled at welding, you can custom-design your own.

An aluminum gate is lighter than wood or steel, so it puts less weight on the gate post. Aluminum weathers better than both wood and steel but is more easily damaged by pressure or abuse.

A stock panel makes a relatively inexpensive, light-duty gate. Unless the panel is framed with wood or steel, though, it will soon warp and sag.

Metal gates are most commonly used for wire fences. If the fence is electric, take care to isolate the gate from hot wires so you won't get shocked going through. If the fence is not electrified and you frequently drive stock through the gate, keep the animals from crowding or rubbing against the wire by nailing wood rails on both sides between the gate posts and the first line posts.

Hinges

Metal hinges, latches, and other gate hardware should be galvanized to resist corrosion. The kind of hinges you use will depend on the type and weight of your gate and on whether you want it to swing both ways. A drive-through gate that swings either way is convenient; you can pull right up to it without worrying that your vehicle will interfere with opening the gate. However, if your fence follows a slope, you will have to hinge the gate to open downslope.

Hinges come in two main styles, standard or eye-and-pin. Standard hinges, suitable for wood gates and square posts, open only one way. They consist of a pair of metal plates connected with a pin that is either removable or permanently fixed. One plate screws to the post, the other to the gate.

The size of the plates determines how heavy a gate the hinge will hold. Butt hinges, the familiar kind used for household doors, have very narrow

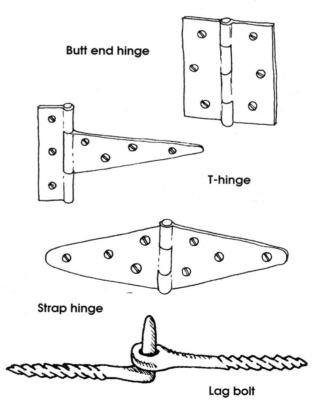

Butt end hinge

T-hinge

Strap hinge

Lag bolt

plates suitable for lightweight walk-through gates. Strap hinges have long rectangular or triangular metal plates that distribute the weight of a heavy gate over a greater area.

For a gate of medium weight, you can get by with a combination or T-hinge. The side that screws to the gate is shaped like a strap hinge; the side that screws to the post is shaped like a butt hinge.

Attach any of these hinges with the largest screws that won't poke through the gate. Attach the gate side first, then the post side. Set a stop-board on the latch post so the gate won't swing beyond it and pull the hinges loose.

To make budget hinges for a lightweight gate, cut rectangular pieces of rubber from a recycled car tire and screw them to the gate and post. Face the convex side down and these hinges will serve as springs to keep the gate closed. Face the convex side outward and the gate will spring open when you release the latch.

For a very heavy gate, a metal gate, or a gate designed to swing both ways, you will need eye-and-pin hinges. The pin side consists of either an L-shaped lag bolt, to be threaded through a wood post, or a short rod fixed to a strap that encircles a pipe post. The gate side of the hinge has an eye into which the pin slips.

Orient the top pin downward so the gate can't be lifted off its hinges. Attach the gate side hinges loosely and slip them onto the pins. Make any necessary adjustments so the gate swings freely and its top lines up with the top of the fence, then tighten the hinges securely.

If you install eye-and-pin hinges parallel to the fence line, you can swing the gate either way but you won't be able to push it flush against the fence on either side. If you want to open the gate all the way on one side, rotate the pin hinge 45° toward that side.

Latches

Latches come in every size, style, and description ranging from a simple coil of baling wire that slips over the gate post to a bolt-action lock. Select a style that can't be opened by wind or clever animals. To discourage vandals and thieves, get a latch that locks. Latch a single gate to the latch post; latch the swinging side of a double gate to the anchored side.

As a quick way to make certain your gate is latched at night, attach reflectors where you can see them only when the gate is closed. A good reflector picks up a flashlight beam from a quarter mile (400 m) away, letting you make your nightly gate check without doing a lot of walking.

A Well-Hung Gate

A gate that supports its own weight is easier to operate and lasts longer than one that drags on the ground. Your gate won't sag if it is properly constructed, the gate post is firm, and the gate is hung plumb so it swings freely. If you don't have someone to hold the gate while you adjust its hinges, support it with blocks underneath to get it just the

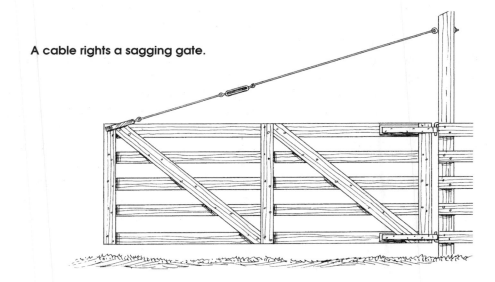

A cable rights a sagging gate.

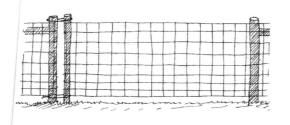
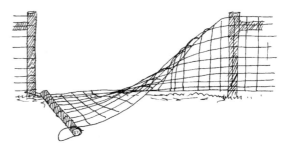

Frameless gate closed (left) and open (right)

Slip-rail gate

right height from the ground.

If the gate sags despite your best efforts, there are several things you can do. If the hinge post doesn't lean, add a third hinge or substitute larger ones. Or run a diagonal brace wire from the top corner of the gate's hinge side to midway down the latch side if it's a walk-through gate, to the bottom corner if it's a drive-through gate. Add a turnbuckle to adjust the tension.

If the hinge post leans, substitute a larger post or beef up the bracing. Or add a support post 8 inches (20 cm) down the fence, wedge a pressure-treated 2x4 (5x10 cm) or 4x4 (10x10 cm) brace block at ground level between the two posts, and use an extra-long strap hinge to tie the two posts together.

Alternatively, increase the height of the hinge post and run a brace cable from the top of the post to the top corner of the gate on the latch side; tighten the cable with a clevis or turnbuckle. If your land is level, rig a small wheel, recycled from an old wheelbarrow, to the bottom of the gate's latch side.

When all else fails, rip out the post assemblies and start again. There is nothing worse than wrestling a balky gate after a hard day's work.

FRAMELESS GATES

A frameless gate, also called a "Western" or "poor man's" gate, makes a fast temporary gate until you

can install a proper one. It is also suitable for an entry that is seldom used. It is awkward, however, if you have to go through it often, especially in bad weather. This gate isn't rigid, so it tends to tangle and snag. The wider the gate, the more likely it is to tangle.

The gate consists of wire strands or a length of woven wire fastened between one gate post and a short, movable pole. The pole is attached to the second gate post by means of two loops of smooth wire at the top and bottom.

To open the gate, slip the top loop up, lift the pole out of the bottom loop, and lay the gate on the ground. Take care not to turn the gate end for end or it will tangle. If you are on foot, leave the pole in the bottom loop while you step through.

To close the gate, set the pole into the bottom loop, push the pole until it meets the gate post, and slip the top loop over. The bottom loop is always attached to the gate post. The top loop is often attached to the post as well, but a taut gate is easier to work if the top loop is attached to the pole instead.

To aid in pulling the gate wires tight enough to prevent sag, as well as getting an animal-tight fit between the post and pole, wire a wooden stick to the gate post and use it as a lever to push the pole against the post. Or use an old seat belt as a gate tightener.

Slip-Rail Gate

A slip-rail gate is a frameless gate for a post-and-rail fence. A pair of posts on each side holds rails across the gateway. To open this gate, simply slip the rails out of the way.

Electric Gates

An electrified gate is a frameless gate consisting of hot wires strung across the gateway. Choose any conductive material — high- or low-tensile steel wire or electroplastic tape or twine — using the same number of strands the fence has.

Each strand is individually opened and closed by means of an insulated handle with a hook on the end. The handle hooks either into a loop of hot wire attached to an insulator on the gate post or into an insulated, energized plate designed for the purpose.

As long as the handle is in place, the electrical connection remains unbroken. As the hook gets rusty, though, less and less current will get through. Eventually you will have to replace the handle.

Some handles are spring-loaded to keep the gate wires from sagging. If you use handles without built-in springs, reduce sag with lightweight expansion springs.

As an alternative, get a "spring" or "slinky-style" gate consisting of one big expansion spring with an insulated handle at one end. The spring expands to fit any normal-size gateway and contracts when the gate is opened. A spring gate requires less muscle than a simple wire strand, which must be pulled taut. On the down side, if you use several springs

across the same gateway, they will tend to tangle when you open the gate.

An option designed to eliminate tangling is the so-called "roller" gate consisting of independent parallel strands that retract into a tube, like retractable clothesline. To close the gate, pull the tube across the gateway and attach it vertically to the gate post.

If you regularly herd livestock through the gate, think twice about electrifying it. Once animals learn they will get a shock at the closed gate, you will have a hard time convincing them to go through when the gate is opened. For that reason, you might prefer a nonelectrified framed gate. If necessary for predator control, run a scare wire along the outside of the gate.

Drive-Through Gate

If you only occasionally need to drive through an electric fence, you may not need a gate at all. Just shut off the energizer, slip the line wires from their insulators, and have someone hold the wires close to the ground while the vehicle passes over. But watch out for protruding parts that might hook and snap the fence wires.

For a gateway you drive through frequently, get an electric drive-through gate that lets you enter from either side. Arms extending from the two gate posts meet in the middle of the gateway. Your vehicle simply pushes between the arms as it passes through, its rubber tires insulating you from the ground to prevent shock. The gate supposedly won't mar your vehicle's finish, but don't bet your new Mercedes on it.

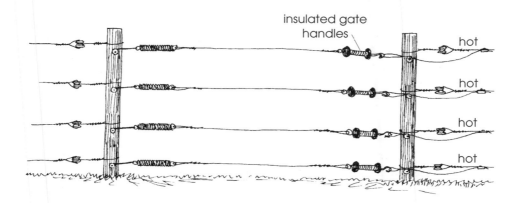

Frameless electric gate

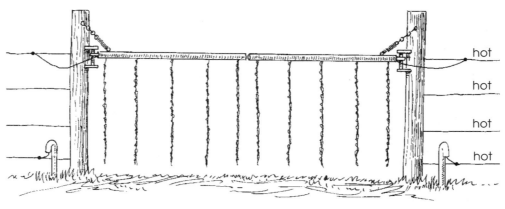

Drive-through gate

You can safely walk through, as well, since the two arms are insulated. Push them open to get through and they will automatically swing shut.

Electrified streamers hanging from the arms discourage cattle, horses, sheep, goats, pigs, and turkeys from wandering through. The streamers get tangled, though, and will eventually break off, rendering the gate ineffective. One version comes without streamers, but works solely for cattle.

Underground Connection

An electric frameless gate should be connected so that the current pulses through it only when it's closed. If it remains energized when it's lying on the ground, it will put extra strain on the energizer. To keep your fence fully energized on both sides of the gate, even when it is opened, run feeder wire beneath the gateway.

An underground connection is a good idea in any case, because gate handles don't provide strong electrical contact. If you install a nonelectrified gate, you will definitely need an underground feeder unless you locate your energizer where it powers the fence on both sides of the gate.

Running the connection underground is better than stringing it overhead because the feeder wire can't be seen and therefore is not unsightly. And it won't attract lightning or be damaged by wind or equipment passing below.

If you have a wire-return system, bury a ground-wire feeder as well as a hot-wire feeder. Use green and red insulated cable designed for burial, 12½-gauge or larger. For a drive-through gate, protect the cables with ½-inch (13 mm) PVC pipe, insulated

tubing, or recycled garden hose. To prevent rain water from collecting in the tubing and shorting the wire, orient the ends downward.

ALARM SYSTEMS

One of the hazards of living in the country is finding strangers wandering around your property. Yet keeping the front gate locked isn't always convenient. Instead, install a driveway alarm system.

There are two options. One is an inexpensive infrared entry alarm, a light-based alarm, that detects both people on foot and moving vehicles. But the light can easily be seen by nighttime trespassers, who might go around it, and the alarm can be tripped by wandering pets and wildlife.

The other option is a buried sensor that detects moving vehicles but not walking intruders. When a car or truck drives by, the sensor triggers a buzzer or burglar alarm, turns on a floodlight, or rings your doorbell.

Bury the sensor 6 inches (15 cm) deep along the shoulder of your driveway and connect it by means of an underground cable to its electric or battery-powered control panel. In the unlikely event that your driveway is more than 10 feet (3 m) wide, place the sensor in the middle.

Fence Alarm

For a woven wire fence, including chain link, you can get an intrusion detection system consisting of a series of sensors mounted on every other fence post and connected to a horizontal cable attached midway up the fence. If the sensors pick up motion

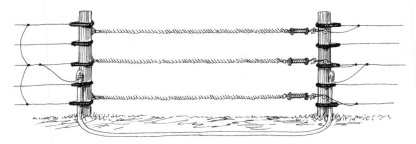

Gate electrifed

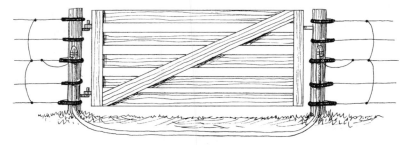

Gate not electrified

caused by climbing over, crawling under, or cutting through the fence, they set off an alarm. Such a system isn't cheap, but it's a surefire way to protect valuable plants or livestock.

If you have an AC-powered electric fence, you can install an alarm that will detect a drop in voltage if someone opens your electrified gate or snips your fence wires to rustle stock or roam your pastures in an off-road vehicle. The same device, installed at the end of your fence line, will warn you of energy leaks in the system. Some models will set off a pocket beeper as well as an alarm. Some have adjustable voltage control so the alarm won't sound for minor or temporary drains. In units with preset voltage, rain or dew or blowing grass can cause a false alarm.

LAZY GATES

Another security device is a remote control or lazy gate. It is operated, like a garage door, from inside your vehicle, house, or barn. One version opens when you beep your horn (which can be a problem if you have friends who toot when they pass by). Some models close automatically after giving you time to drive through. Some swing the gate upward, instead of sideways, so you don't need a large

turn-out to get off the right-of-way.

Remote gate operators can be powered by household current, by battery, or by the sun. Solar models are unsuitable in cloudy climates, since they store only enough energy for about two dozen gate openings.

Optional accessories include push-button, key-switch, and digital key pads. A push-button pad lets you find out, through an intercom, who is at the gate before you open it. A key-switch or digital key pad lets you distribute keys or assigned code numbers to guests or hired hands who need to get in and out.

STILES

A stile is a structure that lets you get to the other side of a fence without opening a gate. It offers an alternative to climbing the fence which, if done repeatedly, eventually causes damage.

Stiles are suitable for occasional use only. You certainly wouldn't want to negotiate one every day while carrying feed or water. Stiles are appropriate only around large or clumsy livestock. If your fence is built to deter clever creatures like dogs, coyotes, or goats, forget the stile.

Stiles come in two basic designs: stepladder and walk-through. Stepladder stiles, as you might ex-

Stepladder stile

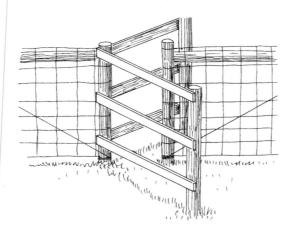

Zigzag stile

pect, must be climbed over. A walk-through requires an opening in the fence with anchor posts on either side. If your fence is not highly tensioned, instead of anchor posts you can set extra-long line posts with a horizontal overhead brace to take up pull.

The most successful walk-through is the zigzag stile, also called a "V-passage" or "English gate." You have to turn your body to get through, making this the original turnstile. It is designed to deter horses and cattle, neither of which can turn handily. If you make the stile very narrow, it will also work with grown sheep. Add a 20-inch (50 cm) step-over sill, and you can use it with hogs.

Construct any stile of pressure-treated 2x4s (5x10 cm) except where platform stairs are involved, then use 2x6s (5x15 cm). Put the stile together with large

ring-shank or screw-shank nails or, even better, lag bolts.

Anchor a stepladder stile to a line post, drive its legs into the soil, or make sure the structure is sturdy and sits squarely on the ground. Provide a handhold at the top. If your fence is electric, slip a length of insulated tubing over the highest wire to prevent nasty surprises.

CATTLE GUARDS

A cattle guard, like a stile, is basically a gate that is always open. Unlike a stile, which admits only foot traffic, a cattle guard handles vehicles as well as people. It is the perfect substitute for a gate where you frequently drive back and forth or where a public right-of-way crosses your land.

A permanent cattle guard consists of a grate covering a drainage pit, designed to discourage animals from crossing by providing unsure footing. The pit also increases the guard's longevity by reducing moisture contact between the grate and the ground. The grate, of course, must be strong enough to hold the weight of heavy vehicles.

You can order a cattle guard ready-made and even have it installed — all you have to do is dig the pit. Commercially made guards are constructed of concrete, galvanized pipe, or steel tubing. Precast concrete won't bend in the middle or come loose at the ends, as pipes or tubing will. If you opt for tubing, round tubes are stronger than square and won't give animals as much foothold.

Construction plans for homemade cattle guards are available from many county extension agents and land-grant universities. Homemade guards are commonly made of pressure-treated lumber; they rarely last as long as a properly built commercial one. Alternative materials for constructing the grate include steel rails, small I-beams, or 2- to 3-inch (50 to 75 mm) galvanized pipes.

A standard guard consists of an 8-foot-wide (2.5 m), 2-foot-deep (60 cm) pit, 10 to 16 feet (3 to 5 m) long. To control weeds, line the pit with concrete, heavy planks, or construction plastic weighted down with rocks. Grade the surface area away from the pit on all sides so runoff water will neither deposit silt in the pit nor collect and create a breeding ground for mosquitoes.

Cover the pit with a grate consisting of parallel

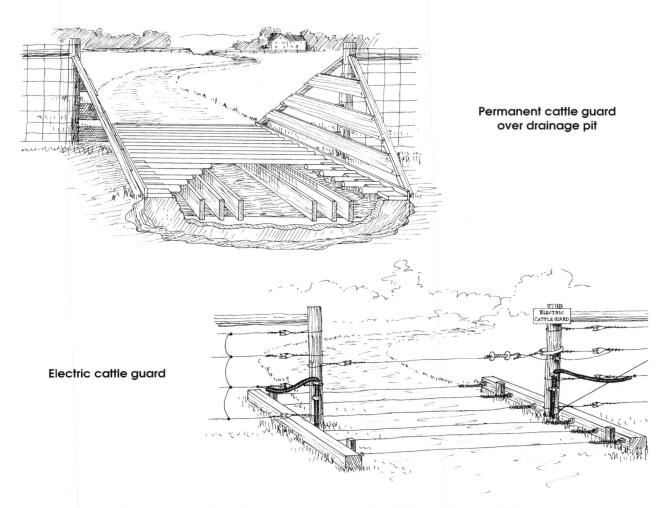

**Permanent cattle guard
over drainage pit**

Electric cattle guard

pipes or heavy planks set on edge. Space them no more than 3 inches (75 mm) apart to minimize jolting to vehicles driving across. Run the pipes or planks perpendicular to the roadway, making them long enough to admit the widest vehicle that's likely to pass.

Slanted triangular end panels, or shields, are often used to reduce the required width and still let large vehicles across. If you will need to drive stock over the guard, make rectangular shields of solid planking that can be lowered to cover the grate. As an alternative, put a gate next to the guard to herd animals through.

If your fence is electric, you might be tempted to put in an electric cattle guard consisting of parallel electrified wires run close to the ground. An electric guard is easy to install and fits any size opening, but works well only on level land. Otherwise, low-slung vehicles can hook a wire and drag the whole thing away.

In addition, animals will learn to step between the wires unless a nose-high hot wire is installed across the opening for a day or two. Even when the wire is down, animals will have learned to associate that area of the fence with pain. As a precaution against predators and adventurous stock, put the hot wire back up at night.

For a temporary situation, use a movable guard that rests on the road's surface. Make it of wood or buy one with flexible plastic rails fastened to a metal frame.

Never use a cattle guard at the entry to a corral since animals may crowd each other onto it. And be cautious about using one for nonbovine stock. Horses may injure their legs. Sheep and goats, with their dainty feet, can easily dance across. Pigs and some sheep *may* be deterred if you dig the pit extra deep. In the final analysis, there is a good reason these are called "cattle" guards.

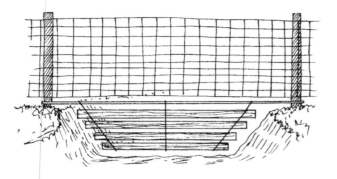

pressure-treated wood

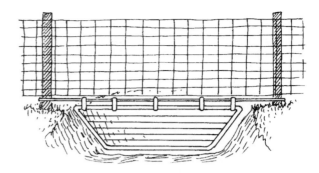

wood or welded steel frame

Water gates

WATERGATES

A watergate, also called a "floodgate," "watergap," "floodgap," or "skirt," crosses a stream or gully, closing the gap beneath the fence section while letting water and floating debris — but not animals — get through. Most gates are free-swinging so driftwood and other trash can't pile up against them. Otherwise, water may back up and break the gate.

Standard fencing is usually stretched across the stream or gully above the gate, but sometimes the gate itself doubles as a section of fence. It is generally hinged by means of metal straps or loops of high-tensile wire and is hung from either a strong cable or two strands of high-tensile wire, twisted together.

Stretch the cable between the bottoms of posts on either side of the wash and tighten it with a turnbuckle or in-line tensioner. If erosion is heavy or gushing water is likely to push hard against the gate, install anchor assemblies on both sides or secure the cable to the next posts in line.

Design the gate to conform to the size and shape of the span it must cross. Make it of pressure-treated wood, woven wire, steel pipe, or sheets of galvanized steel. If the span is greater than 20 feet (6 m), divide the gate into a series of panels that swing individually.

For a wooden gate, fasten horizontal 1x4s (2.5x10 cm) together with Class 3 galvanized, 9-gauge soft wire. Staple the wire to each board and twist it between the boards so they are spaced 4 to 6 inches (10 to 15 cm) apart.

Alternatively, fashion a frame of wood or welded steel and fill it in with rails, woven wire, or parallel high-tensile wires. Or trim a sheet of galvanized metal to fit the gap.

Instead of a swinging gate, you might construct a fixed incline gate. When water rises, debris washes over the top. Wood poles make an inexpensive gate, although one you will have to repair or replace as the poles rot. For a permanent structure that will never rot or wash out, use 2-inch (5 cm) galvanized steel pipes set in concrete.

Space the poles or pipes 12 inches (30 cm) apart for cattle and other large stock, 6 inches (15 cm) apart for sheep and goats. Set them at a 60° angle, pointing toward the direction of water flow as well as toward the stock side of the fence (which presents a problem if the water flows the wrong way).

No matter what kind of watergate you build, after every heavy rain you will have to check it and make necessary repairs. A swinging gate won't always free itself entirely of debris, but may instead jam open until you come along and clear it out.

Electric Watergates

A freely swinging watergate offers clever livestock and predators a way to get through your fence. If the gate is electrified, though, animals will be reluctant to nose through.

Run scare wires across the watergate or, if the gate is made of metal, electrify the whole thing by hinging it to a hot wire in your fence. Alternatively, construct an electric curtain.

Introduced from New Zealand in 1980, an electric curtain consists of lengths of 9-gauge galvanized chain suspended from electrified high-tensile wire so they reach 6 inches (15 cm) above the ground or water. Space the chains 6 inches (15 cm) apart for sheep and goats, 12 inches (30 cm) apart for cattle.

When rising water submerges an electric curtain, or any electrified watergate, it drains current from the rest of the fence. For that reason, it is a good idea to insulate the gate's support cable and energize it separately by means of a watergate controller (available from any electric fence supplier).

Also called a "floodgate isolator" or an "energy limiter," this little device shuts off when it senses a loss of current caused by rising water. As soon as the water recedes, it automatically goes back on.

The isolator normally operates at a lower voltage than the rest of the fence, so it works well for horses and cattle but not for sheep, goats, dogs, and coyotes. If your fence is designed to control any of these critters, in place of a controller install a cut-off switch between the floodgate and the rest of the fence and shut off the current manually during serious flooding.

As an alternative to any of these methods, construct a run of temporary electric fencing across the water gap. During serious flooding, the section will break or wash out without damaging the rest of the fence.

Split gates for greater than 20 feet.

pressure-treated wood

stock panel with pressure-treated wood weights

Fixed incline gate

wood poles or steel pipes

flow

GLOSSARY

▼▲▼▲▼▲▼▲▼▲▼▲▼▲▼▲▼▲▼▲▼▲▼▲▼

Amperage — the amount of energy flowing through an electrified line wire, measured in amperes

Amps — abbreviation for "amperes," a measure of the amount of current flowing through an electric circuit, determining the severity of shock

Anchor and brace assembly — the reinforcement at an end post, corner post, or gate post to prevent fence tension from loosening the post or pulling it out of the ground

Anchor plate — a small sheet of metal attached to the bottom of a post to increase soil contact

Anchor post — any end, corner, gate, or other braced post

Angle post — a corner post

Annealed wire — metal wire that has been heat treated to make it less brittle

Apron fence — netting with a flange at the bottom to discourage digging; also called "beagle netting"

Auger — a post hole digger that operates by turning or twisting; sometimes called a "digger"

Barbed wire — two strands of smooth wire, twisted together and fitted with regularly spaced spikes

Barbless wire — two strands of smooth wire twisted together like barbed wire, but without the barbs; also called "twisted barbless cable"

Batten — a spacer

Bedlog — a deadman

Boundary fence — any fence running along the property line; also called a "perimeter fence"

Boxed end assembly — an H-brace assembly

Brace assembly post — anchor post

Breast log — a deadman

Bright staples — nongalvanized fence staples

Brown back — searing of vegetation touching an electric fence wire

Buck fence — a wood fence built entirely above ground; also called a "jackleg" fence

Cable jack — a come-along

Canadian fence — wood slats spaced vertically and held together with twisted wire

Chain link — continuous mesh steel wire, woven without knots except at the selvage; also called "fabric"

Charger — an energizer

Circuit — the path along which electrical current flows

Clevis — a wire tightening device consisting of half a turnbuckle

Closed circuit — complete electrical circuit through which current flows from the controller through the fence, through the grounding system, and back to the controller

Collector fence — any fence designed to collect blowing sand or snow

Come-along — a ratchet wire stretcher

Compression brace — an H-brace

Compression spring — a coil of wire that contracts when pulled

Concentric galvanizing — zinc coating evenly distributed over the surface of steel wire

Conductance — the ability of a material to allow an electrical pulse to flow through it

Conductor — any material through which electrical energy readily flows

Controller — an energizer

Conventional controller — a high-impedance energizer

Conventional fence — a low-tension fence consisting of soft wire, closely spaced posts, and a conventional controller

Copperclad — copper coated (not the same as "coppered wire," which is unsuitable for fencing)

Corner post — a post at which the fence abruptly changes direction

Crib — to chew

Cross-fence — a fence used to divide an area into smaller portions, so-called because it crosses the land rather than running around the perimeter

Current — flow of electricity

Curve post — a post at which the fence gently changes direction

Cut-out switch — a device for shutting off current to all or part of an electric fence

Dancer — a spacer that doesn't touch the ground

Dead-end splice — an end splice

Deadman — a heavy, bulky item such as a steel plate, chunk of pressure-treated wood, or hunk of concrete buried at the foot of a post as bracing; guy wire anchor

Dead short — serious energy leakage from an electric fence such that little or no current gets through

Deflector fence — any fence designed to divert wind or blowing sand or snow

Dielectric — an insulator

Digger — a tool for making post holes

Dip post — a line post located at a low point in the terrain

Dog leg — a jog in a fence line

Drawknife — a tool used to remove bark from a log

Drivecap — steel or high-impact plastic cover used to protect the top of a fiberglass post as it's driven into the ground with a hammer

Driver — a tool for driving posts into the ground

Dropper — a spacer

Earthing — grounding

Earth monitor light — a feature of some energizers that tells you the system is properly grounded

Earth-return system — an electric fence that relies on soil conductivity to complete the circuit

Earth stakes — grounding rods

Elastic limit — the yield point of wire, approximately equal to 75 percent of its breaking strength

Electric curtain — a watergate consisting of vertically hanging electrified chains

Electrolysis — corrosion caused when dissimilar metals come into contact in the presence of moisture

Electroplastic — twisted or woven strands of plastic shot through with metal filaments capable of carrying electrical current

End post — a post, usually at an opening or gate, from which the fence travels in one direction only

End splice — wire tied off at an anchor post or in-line tensioner; also called a "dead-end" or "eye" splice

Energizer — the unit that gives an electric fence its jolt; also called a "charger," "controller," or "fencer"

Energy limiter — a watergate controller

Eye splice — an end splice

Fabricated fence — woven wire

Fascia — a board nailed onto the posts of a post-and-plank fence, sandwiching the rails; a board nailed to the top of a post-and-plank fence to increase its strength

Feeder — a wire that transmits electrical current from one wire to another; also called a "jumper"

Fence jack — a wire stretcher

Fence pliers — an all-purpose fencing tool; also called "hammerhead" pliers

Fencer — an energizer; a person who builds fences

Field fence — woven wire with relatively wide wire spacings

Filler wires — all the horizontal wires of a woven fence between the top and bottom line wires; also called "intermediate line wires"

Floodgap controller — a watergate controller

Floodgate — a watergate

Foil trainers — strips of aluminum foil tied to an electric fence to attract animals and ensure they get zapped

Foots — wedge-shaped blocks used to add stability to an anchor post

Fuse — a safety device that prevents the flow of too much current through an electrical circuit

Galvanized — coated with zinc

Game fence — a fence designed to control domesticated herds of wildlife

Gauge — a number indicating the thickness of wire — the lower the number, the thicker the wire

Grade — the quality of wood based on the presence or absence of knots, checks, decay, and other defects; ground level; to level the soil

Grounded wire — a wire that is connected, directly or indirectly, to the earth

Grounding — connecting an electric fence system to the earth to complete the circuit

Guy — a support wire; also called a "stay"

H-brace — horizontal brace between two posts; also called a "boxed assembly" or "post-and-rail assembly"

Heartwood — dark wood that extends from the pith (center) of a log to the sapwood and is more resistant to decay than sapwood

Helical wire grip — twist connector

High-tensile wire — line wire with tremendous breaking strength

Horizontals — wires running lengthwise along a fence line; also called "line wires"

Hot — electrified; also called "live"

Hot tape — electroplastic tape

Impedance — measure of how easily current flows through an electrical circuit

Induction — the transfer of power without direct contact, such as the charging of a neutral line wire by a parallel hot wire

Induction coil — several loops of wire that set up a magnetic field to impede the flow of electric current

In-line tensioner — a permanently installed device used to tighten fence wire; also called "in-line tighteners"

Insulator — any material that resists the flow of electricity

Iowa wire — four-point barbed wire

Joule — the amount of energy needed to produce one watt for one second

Knuckled — characteristic of the selvage edges of chain link fabric in which adjoining ends are interlocked and then bent back to form a closed loop

Knurling — barbs on a fence post staple, designed to increase its holding power

Lead-out wire — insulated wire connecting an energizer to the fence and to ground rods

Leak — drainage of electric current from a live wire to ground, causing voltage to drop

Lightning arrestor — a lightning diverter

Lightning brake — an induction coil used to increase resistance in electrified line wire so a lightning strike will be encouraged to seek an alternate route

Lightning diverter — a device that channels high-powered electrical energy to ground

Line post — any post not in an anchor position

Line wire — any horizontal wire or cable running the length of the fence line

Live — electrified; also called "hot"

Living fence — a fence consisting of trees, shrubs, or other growing plants

Lock — the angles where the rails of a zigzag fence meet

Mains — plug-in energizer

Merchant wire — soft, smooth wire

Mortise — a slot in a wooden post to hold rails

Negative — grounded

Netting — high-gauge soft wire woven into small mesh

Neutral — neither energized nor earthed

Neutral return — the ground of an electric fence

Offset wire — scare wire

Off-time — the interval between electrical pulses produced by an energizer

On-time — the interval during which an energizer produces an electrical pulse

Open circuit — an incomplete electrical circuit, for example one caused by broken wires

Outrigger — scare wire

Peanut butter fence — an electric fence with peanut butter smeared on foil trainers (or electroplastic tape) to attract wildlife

Peeling spud — a drawknife

Perimeter fence — a fence running along the property line; also called a "boundary fence"

Physical barrier fence — any fence that keeps animals in or out solely by virtue of its strength

Pickets — verticals in woven wire; spacers

Pinlock insulator — a device, used to isolate electrified wire, that holds the wire by means of an inserted pin

Plumb — straight and vertical

Polarity — the condition of electric fence wire as being either energized or earthed

Poly tape — electroplastic tape

Poly wire — electroplastic wire

Porosity — the percentage of open space to the total surface area of a deflector fence

Positive — energized

Power fence — electric fence, usually one controlled by a high-energy, low-impedance energizer

Prefabricated fencing — woven wire

Preservative — chemicals used to protect wood from deterioration as a result of decay, insects, marine borers, weathering, and water absorption

Pressure — stress caused by pushy animals; in pounds, a measure of line wire tension; a method for infusing wood with preservative

Pull-out — the tendency of a staple to come out of a wood post; the tendency of a post to lift out of the ground; the horizontal distance between vertical stays in woven wire

Pull-post — a braced line post

Pulse — a short burst of electrical energy

Pulsed wire — a fence wire carrying intermittent current; also called a "live," "hot," or "positive" wire

Rammer — a tamper

Rebar — steel rod used to reinforce concrete

Resistance — opposition to the flow of electricity, measured in ohms

Ring fence — a boundary fence

Rise post — a line post located at a high point in the terrain

Rod — measurement (equal to 16.5 feet or 5 meters) used by surveyors and fence wire manufacturers; a solid fiberglass post; rebar

Run — length of fence between two anchor posts; also called a "strain"

Running post — line post

Sacrificial action — corrosion of a wire's zinc or aluminum coating instead of corrosion of the wire itself; also called "sacrificial corrosion"

Sand fence — a fence designed to deflect or collect blowing sand

Sapwood — the newer, light-colored wood between the heartwood and the bark of a log, which decays more rapidly than heartwood

Selvage — strands of smooth wire at the top or bottom of a woven fence that are added to increase strength

Selvage wire — the top and bottom horizontal in woven wire or netting

Short circuit — serious energy leakage; called a "short" for short

Short-shock system — an electric fence controlled by a high-energy, low-impedance energizer

Sighter wire — any particularly visible wire used to make an electric fence easier to see

Skirt — a watergate

Snow fence — a fence designed to deflect or collect blowing snow

Soft wire — line wire with relatively low breaking strength

Spacer — a rigid or semirigid rod or stiff wire used between line posts to maintain wire spacing; also called a "stay," "batten," "dropper," "dancer," or "picket"

Span — the distance between two posts; a brace

Sphere of intimidation — the space around an electrified line wire which an animal won't enter willingly

Spinning jenny — a dispenser for smooth wire

Standard — a fence post

Standoff or standoff bracket — a scare wire insulator

Stay — a spacer; the brace in an anchor assembly; a brace wire; a guy wire

Stepped — a technique for installing rail fence on sloping terrain, whereby each set of rails is lower (or higher) than the preceding one, creating a stair-like effect; also called "step down"

Stock (as in stock fence or stock panel) — livestock

Strain — a straight run of fence between anchor assemblies; to tension wire

Strainer — a tool for tensioning wire

Strainer assembly — anchor assembly

Strainer post — an anchor post

Stretch brace — a pull post assembly

Stretcher post — a braced line post; also called a "pull post"

Strip fence — an unconnected length of fence used, for example, along a woodlot to keep deer away from crops

Surge — sudden increase in the flow of current through an electrical a circuit; short for "power surge"

Surge suppressor — an induction coil or other device used to reduce sudden high voltage due to a lightning strike or sudden burst of power in the transmission line

Suspension fence — a fence with its posts so far apart that the fencing material seems to hang suspended

Swedged-end rails — pipe rails having one female end and one male end for easy joining

Swedish fence — an old-time wooden snow fence

Tamper — a heavy bar used to compact soil around a hand-set post; also called a "rammer"

Tensile strength — the amount of pulling force (in pounds per square inch) needed to break a line wire

Tension curves — U-shaped crimps in line wires that serve as tensioning springs

Tensioner — a device for tightening line wire; also called a "tightener" or "strainer"

Tension spring — a coil of wire that expands when pulled

Terminal posts — the end, corner, and gate posts of a chain link fence

Thread-lay wound — a characteristic of coiled wire whereby it unwinds from the outside rather than from the top, an important feature if you are machine-feeding several wires at a time

Three-dimensional fence — a fence deliberately built at a slant

T-post — a post having a cross-section in the shape of the letter T

Trip wire — offset wire

Turnbuckle — a wire-tightening device consisting of two eyebolts threaded into a rectangular frame

Twist connector — a spiral of stiff wire used to splice smooth wire; also called a "twist link," "helical wire grip," "vine line connector," or "wrap connector"

Twisted — characteristic of the selvage ends of a chain link fence in which adjoining wires are twisted together and left sticking out in sharp points

Twitch — to tension diagonal brace wires by twisting

Twitch stick — a short length of wood used to tension brace wires by twisting them together

Two-dimensional fence — two parallel fences with a narrow alley between them; an electric fence with scare wires

Voltage — the electrical pressure of current flowing through a line wire, measured in volts

Voltage spike — a surge suppressor

Volts — a measure of the amount of pressure behind an electrical pulse

Wane — a strip of bark or other defect on the edge of a board or rail

Watergate — a structure for fencing across a gully or stream; also called a "watergap," "floodgate," "floodgap," or "skirt"

Watergate controller — waterproof resistor that reduces the flow of current through an electrified watergate; also called an "energy limiter" or "floodgap isolator"

Wattage — the amount of work done by an electrical pulse flowing through a line wire, measured in watts

Weed chopper — an energizer that burns back vegetation; also called a "weed burner" or "weed clipper"

Welded mesh — woven wire welded instead of knotted together

White rust — chalky white substance indicating that galvanized wire is starting to corrode

Wire grip — the jaws of a wire stretcher; a clamp that holds wire tighter the more you pull

Wire-return system — an electric fence containing both energized and grounded line wires

Wire spinner — spinning jenny

Worm fence — a zigzag fence

Wrap connector — a twist connector

Yield point — the point beyond which, if fence wire is stretched, it will not return to its original length

Zigzag fence — a wood fence in which each section is at an angle to its adjoining sections

CONVERSIONS

▼▲▼▲▼▲▼▲▼▲▼▲▼▲▼▲▼▲▼▲▼▲▼▲▼▲▼▲▼▲▼▲▼▲▼▲

In the United States, standard metric sizes are hard to come by for such things as the lengths of fence posts, the dimensions of milled lumber, and the weights of cement sacks. Throughout this book, I have attempted to make sense of metric conversions by rounding them to reasonable numbers. The nearest metric equivalent of an 8-foot post, for example, is unlikely to be exactly 243.8 cm; thus I have rounded off to the nearest workable measurement. Even so, many conversions may still not make strict sense. I hope that I have provided enough information for readers who work in metric to do any necessary extrapolation. The following conversions are offered as an additional aid.

UNITS OF LENGTH

THIS	equals	THIS
1 link		7.92 inches or 20.1168 centimeters
1 rod		16.5 feet or 5.0292 meters or 25 links
1 chain		66 feet or 20.1168 meters or 4 rods or 100 links
1 furlong		660 feet or 201.2 meters or 40 rods
1 inch		2.54 centimeters
1 foot		12 inches or 0.3048 meters
1 yard		3 feet or 0.9144 meters or 4.5 links
1 mile		5,280 feet or 1.6093 kilometers or 8 furlongs or 320 rods or 80 chains
1 centimeter		0.3937 inch
1 meter		3.2809 feet
1 kilometer		0.6214 mile

UNITS OF TENSION/FORCE

THIS	equals	THIS
1 pound		4.45 Newtons or 0.45 kilograms
1 Newton		.22 pounds or 0.1 kilograms
1 kilogram		9.8 Newtons or 2.2 pounds

SUPPLIERS

▼▲▼▲▼▲▼▲▼▲▼▲▼▲▼▲▼▲▼▲▼▲▼▲▼▲▼▲▼▲▼

Fence supplies are available at most farm stores. Given the vast array of fencing materials, though, you may not always find the materials you want. Some of the best suppliers sell only through mail-order. Many will answer questions and even help you design your fence. If you're not satisfied with the response to your initial inquiry, you may be equally dissatisfied with later customer service — move on to another supplier. This list is offered as a service only, and does not imply endorsement by author or publisher.

Cattle Guards

Easi-Set Industries, Midland, VA 22728, 540-439-8911 (concrete).

Farnam Equipment Company, 1302 Lewross Rd., Council Bluffs, IA 51501, 800-825-2555 (galvanized pipe and flexible plastic rail).

Martin Ranch Supply, Inc., 522 Martin Avenue, Rohnert Park, CA 94928, 707-585-1313 (pipe).

Chain Link

Guardian Wire & Steel, 1103 Pasture Lane, Columbia, SC 29201, 800-845-2526.

New Brunswick Wire Fence, 80 Henri Dunant Street, Moncton, New Brunswick, Canada E1E 1E6, 506-857-8141; 800-561-7986.

PulJak Division, Custom Products Corporation, 457 State Street, North Haven, CT 06473, 800-562-9522 (fabric stretchers).

Electric Fence Supplies

Agricultural Engineering, The Pennsylvania State University, University Park, PA 16802 (high-tensile fence materials list and cost software for Apple IIe).

Common Sense Fence, 13433 Highway 52 Southeast, Chatfield, MN 55923, 800-533-1680 (electrified).

Daken, Box 360, Brookvale, NSW, Australia 2100.

Gallagher Electronics, Ltd., Private Bag, Hamilton, New Zealand 3026 (branches in Australia, Canada, Ireland, and United Kingdom).

Gallagher Power Fence, Box 708900, San Antonio, TX 78270, 800-531-5908.

Gripple, Inc., P.O. Box 368, Batavia, IL 60510, 800-654-0609 (wire splices and tensioning tool).

ITW Fastex/Agtex, 223 E. Imperial Highway #165, Fullerton, CA 92635, 714-566-9150 (Wirevise, Wirelink).

Kencove Fence, 111 Kendall Lane, Blairsville, PA 15717, 412-459-8991.

Kiwi Fence Systems, 1145 E. Roy Furman Hwy., Waynesburg, PA 15370, 412-627-5640.

New Zealand Fence Systems, Box 518, Clackamas, OR 97009, 503-657-3657.

Parker McCrory Manufacturing Company, 2000 Forest Ave., Kansas City, MO 64108, 816-221-2000 (energizers).

Premier Fence Systems, Washington, IA 92353, 319-653-7622.

Twin Mountain Fence, Box 2240, San Angelo, TX 76902, 800-527-0990.

Waterford Corporation, Box 1513, Ft. Collins, CO 80522, 800-525-4952.

West Virginia Fence Corporation, U.S. Route 219, Lindside, WV 24951, 800-356-5458.

Plastic Fencing

B.F. Products, Inc., P.O. Box 61866, Harrisburg, PA 17106, 800-255-8397.

Centaur Fencing Systems, Box 2788, Muscle Shoals, AL 35662, 205-383-0124 (plastic-coated high-tensile wire).

Conwed Plastics, 2810 Weeks Ave. S.E., Minneapolis, MN 55414, 800-426-0149.

Endurance Net, Box 128, Roebling, NJ 08554, 609-499-3450.

J.A. Cissel Mfg. Co., Box 2025, Lakewood, NJ 08701, 800-631-2234.

Tenax Corporation, 8291 Patuxent Range Rd., Laurel, MD 20794, 301-725-5910.

Post-and-Rail

Bufftech, Inc., 2525 Walden Avenue, Buffalo, NY 14225, 716-685-1600 (rigid vinyl).

The Burke-Parsons-Bowly Corporation, Box 231, Ripley, WV 25271, 800-272-2365 (wood).

Centaur Fencing Systems, Box 2788, Muscle Shoals, AL 35662, 205-383-0124 (flexible vinyl).

Common Sense Fence, 13433 Highway 52 Southeast, Chatfield, MN 55923, 800-533-1680 (electrified).

Easi-Set Industries, P.O. Box 300, 5519 Catlett Rd., Midland, VA 22728, 800-547-4045 (concrete).

Faddis Concrete, 3515 Kings Highway, Downingtown, PA 19335, 800-777-7973 (concrete).

Fleming Manufacturing Co., 400 Fleming Avenue, Cuba, MO 65453, 573-885-3311 (concrete).

Kee Industrial Products, Inc., 100 Stradtman St., Buffalo, NY 14206, 800-851-5181 (pipe rail fittings).

Martin Ranch Supply, Inc., 522 Martin Avenue, Rohnert Park, CA 94928 (pipe and prefab pipe panels).

Nebraska Plastics, Inc., Box 45, Cozad, NE 69130, 800-445-2887 (rigid vinyl).

Stroberg Equipment Company, Inc., 603 Urban Drive, Hutchinson, KS 67501, 316-662-7650 (pipe panels).

Triple Crown Fence, Box 360, Milford, IN 46452, 800-365-3625 (rigid vinyl).

Post Hole Diggers (Manual)

Seymour Manufacturing Co., Inc., 500 North Broadway, Seymour, IN 47274, 800-457-1909.

Power Equipment

General Equipment Co., 1500 East Main, Owatonna, MN 55060, 800-533-0524 (portable and hydraulic augers).

Ground Hog, Inc., Box 290, San Bernardino, CA 92402, 800-922-4680 (portable augers).

Hawk Industries, 1245 East 23rd Street, Long Beach, CA 90806, 562-426-9477 (portable post drivers).

Little Beaver, Inc., Box 840, Livingston, TX 77351, 409-327-3121 (augers).

Rhino Tool Company, Box 111, Kewanee, IL 61433, 309-853-5555 (hydraulic post drivers and pullers).

Worksaver, Inc., Box 100, Litchfield, IL 62056, 217-324-5973 (augers, hydraulic post drivers and pullers).

Security Devices

AC Automatic Gate Opener, 1274 Stacks Road, East Ennis, TX 75119, 214-747-8104 (gate openers).

Litton Poly-Scientific, 1213 North Main Street, Blacksburg, VA 24060, 318-925-3064 (intrusion detection systems).

Mier Products, Inc., 1500 North Ann Street, Kokomo, IN 46901, 800-473-0213 (driveway alarms).

Preferred Technologies Group, 2741 B Littiz Pike, Lancaster, PA 17601 (driveway alarms).

Stanley Home Automation, 41700 Garden Brook, Novi, MI 48050, 734-475-3204 (gate openers, driveway alarms).

Winland Electronics, Inc., Box 473, Mankato, MN 56001, 800-635-4269 (driveway alarms).

Snow Fences (See also Plastic Fencing)

New Brunswick Wire Fence, 80 Henri Dunant Street, Moncton, New Brunswick, Canada E1E 1E6, 506-857-8141 (Canadian snow fence).

Rocky Mountain Forest and Range Experiment Station, 222 South 22nd Street, Laramie, WY 82070, 307-745-8924 (printed information).

Tabler & Associates, Box 483, Niwot, CO 80544, 303-652-3921 (design consultants).

Steel Posts

Wedge Loc Company, 1580 N. Pendleton, Rio Rico, AZ 85648, 800-669-7218 (T-post bracing hardware).

Trellis

Calico Supplies, 2060 E. Indiana Avenue, Southern Pines, NC 28387, 910-944-7306 (plastic wire).

Centaur Fencing Systems, Box 2788, Muscle Shoals, AL 35662 (plastic-coated wire).

Vinyl Landscape Trim

People Devices, 686 Richardson, Owosso, MI 48867.

One-Call Services

Free one-call services will help you contact the appropriate utilities in your area to ensure you don't cut into underground pipes or cables when you dig post holes or drive posts into the ground. When you call, ask about the extent and limit of the service offered and write down the name of the person you talk to. In some states, one-call participation is required by law; in others, it is strictly voluntary.

You can obtain the telephone number of your local one-call service by contacting American Public Works Association Utility and Coordination Council One-Call Systems International Committee, 1313 East 60th Street, Chicago, IL 60637, 312-667-2200.

SELECT BIBLIOGRAPHY

▼▲▼▲▼▲▼▲▼▲▼▲▼▲▼▲▼▲▼▲▼▲▼▲▼▲▼▲▼▲▼▲▼▲▼▲▼▲▼

The Best Fences, James FitzGerald, Storey Books Bulletin A-92, North Adams, MA, 1984.

Building an Electric Antipredator Fence, David S. deCalesta, Pacific Northwest Extension Publication PNW 225, 1983.

Building Fences, American Association of Vocational Materials, Athens, GA, 1974.

Excavator's Damage Prevention Guide and One-Call Systems International 1991-1992, American Public Works Association Utility Location and Coordination Council, Chicago, IL.

"Fences for the Farm and Rural Home," Merrill S. Timmins, Jr., USDA Farmers' Bulletin #2247, 1982.

High-Tensile Wire Fencing, Arthur W. Selders and Jay B. McAninch, Northeast Region Agricultural Engineering Service, Cornell University, Ithaca, NY 14853, 1987.

How to Build Fences, United States Steel Corporation, Pittsburgh, PA, 1980 (out of print).

How to Build Orchard and Vineyard Trellises, United States Steel Corporation, 1982 (distributed by Kiwi Fence Systems, Waynesburg, PA).

"Maintenance and Safety for High Tensile Fence," Tara A. Baugher et al., publication 813, West Virginia State University, 1985.

Planning Fences, American Association of Vocational Materials, Athens, GA, 1980.

"Smooth-Wire Tension Fences, Design and Construction," Mark C. Mummert, William L. Kjelgaard, and Zheng Ping Wu, The Pennsylvania State University, College of Agriculture, Cooperative Extension Service, University Park, PA.

"Trouble-Shooting High Tensile Wire Fences," J. B. McAninch et al., publication 821, West Virginia State University, Morgantown, WV, 1990.

INDEX

▼▲▼▲▼▲▼▲▼▲▼▲▼▲▼▲▼

(Illustrations are indicated in *italics;* charts are indicated in **bold.**)